ASSISTED LIVING HOUSING FOR THE ELDERLY

Design Innovations from the United States and Europe

ASSISTED LIVING HOUSING FOR THE ELDERLY

Design Innovations from the United States and Europe

VICTOR REGNIER, AIA

VNR VAN NOSTRAND REINHOLD
New York

To the influential people in my life: my wife, Judy; my daughters, Jennifer and Heather; my mother, Helen; my father, Victor; and my grandmother, Katie.

Library of Congress Catalog Number 92-43118
ISBN 0-442-00702-7

Printed in the United States of America

Van Nostrand Reinhold
115 Fifth Avenue
New York, NY 10003

International Thomson Publishing GmbH
Konigswinterer Str. 518
5300 Bonn 3
Germany

International Thomson Publishing
Berkshire House,168-173
High Holborn, London WC1V 7AA
England

International Thomson Publishing Asia
38 Kim Tian Rd., #0105
Kim Tian Plaza
Singapore 0316

Thomas Nelson Australia
102 Dodds Street
South Melbourne 3205
Victoria, Australia

International Thomson Publishing Japan
Kyowa Building, 3F
2-2-1 Hirakawacho
Chiyada-Ku, Tokyo 102
Japan

Nelson Canada
1120 Birchmount Road
Scarborough, Ontario
M1K 5G4, Canada

16 15 14 13 12 11 10 9 8 7 6 5 4 3 2 1

Library of Congress Cataloging-in-Publication Data
Regnier, Victor, 1947-
Assisted living housing for the elderly : design innovations from the United States and Europe / Victor Regnier.
p. cm.
Includes bibliographical references and index.
ISBN 0-442-00702-7
1. Architecture and the aged—United States. 2. Aged—United States—Dwellings. 3. Architecture and the aged—Europe. 4. Aged—Europe—Dwellings I. Title.
NA2545.A3R45 1993
728'.043--dc20 92-43118
CIP

CONTENTS

PREFACE

INTRODUCTION

It is often said that the value and meaning of a civilization can be determined from the record it leaves in the form of architecture, and that the true measure of the compassion and civility of a society lies in how well it treats its frail older people. Herein lie the two topics this book explores. Both are considered measures of achievement, which should be reaching a zenith in our society as we begin the twenty-first century. Both, however, are considerably underdeveloped, given their transcendent position in society and their timeless nature.

The impetus for this project started years ago and grew from my concern as an architect that nursing homes were not designed to encourage independence or to appear residential in character. The opportunity to carefully study these issues came through an Administration on Aging grant. This allowed me to explore innovations in an emerging housing type called "assisted living," which takes greater responsibility for the health and personal care needs of mentally and physically frail older people in a residential rather than institutional context.

UNITED STATES STUDY METHODOLOGY

Invitations were sent to 1700 facilities, consultants, architects, financiers, and management consultants, asking them to nominate facilities they felt had experimented with innovative ideas in management, financing, and design. One hundred and thirty responses were received from organizations involved in assisted living, board and care, and senior housing.

About six months later, funding was received from the Retirement Research Foundation to pursue an expert nomination process and site visitations. During fall 1990, fifty senior housing knowledgeables were contacted and asked to suggest noteworthy examples of residentially styled facilities providing in-depth personal care services for the very frail. One hundred additional projects were contacted, bringing the total of identified projects to 230. These were reviewed and aspects from projects that were noteworthy identified. Twenty-five projects were selected for site visitations.

The information from these two studies is collected in the manuscript *Best Practices in Assisted Living.* This book is currently being updated, revised, and expanded for Columbia University Press and will appear in print in 1994 as *Assisted Living for the Frail and Aged: Innovations in Design, Management and Financing.*

An important side effect of examining these innovative facilities was a clearer understanding about how rules, regulations, state laws, past experiences, habits, conventional thinking, and lethargy have kept our society from exploring residential alternatives to nursing homes and more aggressively pursuing concepts such as aging in place.

The site visits brought me in contact with a number of people committed to the idea of providing older frail people with residential alternatives to nursing homes. I was impressed by their ability to implement these ideas and avoid bureaucratic tangles that plague this industry. People like Keren Brown Wilson, in Oregon, Paul Klaassen, in Virginia and Alan Black, in Minnesota, were doing the impossible. They were taking care of older frail people in a responsible and professional manner, in comfortable, private, residential settings. Under other circumstances these residents would have been placed in a nursing home. They were providing older residents with dignity, autonomy, independence, privacy, control, choice and in many cases, a lot of health care support.

In the middle of this work, several key conversations with David Hoglund, John Beck, and Jon Pynoos reminded me of the enormous progress northern European countries have made in successfully pursuing a friendlier system of housing and health care support. Most recently, these cultures have experienced a period of institutional reform leading to national policies that abandoned institutions for all except the most medically indigent.

EUROPEAN STUDY METHODOLOGY

During a sabbatical leave in the summer and fall of 1990, I began identifying noteworthy European projects. I started by reviewing the last ten years of published projects from the Avery Index of Architectural Periodicals and the RIBA Index. I found eighty published projects of merit that fit broadly based residential imagery and service provision criteria. Next, seventy-eight letters were sent to architectural periodicals, schools of architecture, professional architectural organizations, environment-behavior researchers, and practicing architects in the five countries I identified (Norway, Sweden, Denmark, Finland, and the Netherlands). I also sent forty-five letters to gerontological research organizations, social service agencies, academic researchers, and health and housing agencies in these five countries. I received fifty replies to these letters, which suggested important contact people and projects of merit. The combination of the published projects and nominations gave me a list of approximately 200 projects from which to choose sites to visit.

I identified the projects on maps and devised a route through these five countries that brought me in contact with the greatest number of settings. Important projects of obvious merit created the structure of the route. However, in my summer schedule, I was somewhat handicapped by the vacation period in Sweden, Norway, and Finland. However, with only a few exceptions, I was able to visit all of the important projects suggested to me by contacts in the architecture and gerontology fields in these countries. The final number of site-visited projects amounted to 100. This included twenty-three from Denmark, twenty-nine from Sweden, twenty-five from the Netherlands, fourteen from Norway, and nine from Finland.

Each project was treated similarly. Typically, I scheduled one site visit in the morning and one in the afternoon, each normally lasting three hours. I prepared a 70 question interview and a 144-item architectural checklist. Both raised critical issues and concerns, which I reviewed at each site. I took approximately sixty photographs at each facility, recording special features and aspects of the project that appeared noteworthy. The typical site visit included a one-hour interview, followed by a tour of the facility and further discussion about the project's important features. Notes taken during the interview process were translated each evening into a "lessons learned" format that captured fifteen to sixty salient items for each project. These were dictated and transcribed as a record of the issues and features identified and reviewed at each project.

The European site visits convinced me of the feasibility of developing community based residential settings for the mentally and physically frail. I visited buildings in Europe that were so radical in their design and service philosophy, they would be considered illegal in the United States. I saw systems that worked better, that offered residents more encouragement and independence, often at costs that were comparable or less than those of private facilities in the United States. It seemed odd that the handful of innovative projects identified in the United States represented the mainstream in Europe. Hundreds of European facilities were serving thousands of older people in situations that provided security, privacy, and a sense of well-being without removing their dignity and sense of independence. Residents with major mental and physical infirmities were able to grow older in housing that was spacious, complete, and tethered to a range of emergency care and supportive health care services. It was a vision I saw only a glimpse of in the United States. In these five northern European countries, it was the norm.

CONCLUSION

Experiences like that change your life and give you special resolve to imagine alternative possibilities. This book represents the best argument I can make to pursue this new way of thinking about how to deinstitutionalize our system and allow people in the United States to grow old and frail with dignity.

Victor Regnier, AIA
Los Angeles, CA.

ACKNOWLEDGMENTS

INTRODUCTION

Research projects like this are always dependent on the help and support of many people. In this case, four major groups were instrumental in bringing about this work. The first is the funding agencies that evaluated proposals and financially supported the research program. The second group includes the academics and professionals who wrote letters of support and helped by making arrangements or identifying specific projects of merit. The third group consists of managers, administrators, architects, and sponsors who hosted me in the United States and Europe, providing insights and impressions about their facilities. The fourth group includes those who took a personal interest in the work by reading the manuscript, talking with me about their beliefs and ideas, and generally providing criticism and encouragement to strengthen weaknesses and clarify ambiguities. This project owes all four groups an enormous debt of gratitude for the important roles they played in its completion.

FUNDING SOURCES

Seven different funding sources provided financial support for the work. They include:

1. Council for International Exchange of Scholars (Fulbright), Washington, DC.
2. The American-Scandinavian Foundation, New York, NY.
3. The Norway-America Association, Oslo, Norway.
4. Health Facilities Research Program, AIA/ACSA Council on Architectural Research, Washington, DC.
5. Fannie Mae Foundation, Washington, DC.
6. National Eldercare Institute on Housing and Supportive Services, Andrus Gerontology Center, U.S.C, Los Angeles, CA (Administration on Aging).
7. Retirement Research Foundation, Chicago, IL.

LETTERS OF SUPPORT

Support letters were necessary to convince funding sources of the merit of the proposed work. These included letters from United States–based architects and gerontologists, as well as letters from the heads of academic institutions, research centers, and service organizations in Europe who agreed to assist me in identifying important places to visit.

Seven people wrote support letters about the need for this work that were compelling. M. Powell Lawton, Ph.D., Director of Behavioral Research, Philadelphia Geriatric Center, Philadelphia, Pennsylvania, not only supported this work, but has been the greatest source of inspiration and encouragement for my work in the last twenty years. James E. Birren, Ph.D., Director, Anna and Harry Borun Center for Gerontological Research, University of California, Los Angeles, and former Dean of the Andrus Gerontology Center at the University of Southern California, as he always has done, took a personal interest in helping me. He shared names from his extensive worldwide collection of colleagues with me, and wrote an inspiring letter of support. Robert Harris, FAIA, Professor and former Dean of the School of Architecture at U.S.C, enthusiastically supported this project as he has supported every one of my previous projects, with unfaltering encouragement and an attitude of caring and intellectual curiosity.

Jon Pynoos, Ph.D., Director of the program of Policy and Services Research at the U.S.C Andrus Gerontology Center, is my colleague and friend. We have helped one another for fifteen years and can tag team lecture, often ending one another's sentences. Jon, as always, was there to support and criticize, making the work clearer and sharper. Neal Cutler, Ph.D., Director of the Boettner Institute of Financial Gerontology at the University of Pennsylvania, in Philadelphia, read the original Fulbright proposal and took it apart in detail. Neal is responsible for the Retirement Research Foundation grant. It was his criticism about the nature of the work that gave me the idea of doing a study, which directly compared the United States and European examples on similar attributes. Sandra Howell, Ph.D., Professor of Behavioral Science, School of Architecture and Planning, MIT, Cambridge, Massachusetts, through conversations and discussions about her own cross-cultural work in Japan, suggested several approaches and directions for study.

Finally, David Hoglund, AIA, Partner with the New York design firm of Perkins Eastman, provided the precedent for this study. His 1985 European study fascinated me, and his extraordinary work with Woodside Place put me in awe of his talent, his humanity, and his resolve. He is a former student who has surpassed all of my expectations.

The contact people who wrote detailed letters of support came from a variety of different places. Each provided a range of insights, pointing out attributes of projects and service systems that were worthy of exploration. They included:

1. Jan Paulsson, Chalmers Institute of Technology, Göteborg, Sweden
2. Stig Berg, Institute for Gerontology, Jönköping, Sweden
3. A.M.E. v.d. Weerd, WVC, The Haag, the Netherlands
4. Peter Houben, Technical University at Delft, the Netherlands
5. Cees Zwinkels, De Architects'Gravenhage, the Netherlands
6. Rikard Kuller, Lund Institute of Technology, Lund, Sweden
7. Jan-Erik Ruth, The Kuntokallio Foundation, Helsinki, Finland
8. Sven Thiberg, Royal Institute of Technology, Stockholm, Sweden
9. Per Rygh, Norske Arkitekters Landsforbund, Oslo, Norway

HOSTS AND SPONSORS

The third group is those who spent time accompanying me to several places, answering questions, and communicating ideas. They alerted me to attributes and qualities of noteworthy examples that were important to comprehend and experience. I could tell stories about each and describe their generosity in many different ways.

1. Carol Fraser Fisk, Executive Director, ALFAA, Oakton, VA
2. Erik Ejlers, Architect, Copenhagen, Denmark
3. Bente Lindstrom, Professor, School of Architecture, Århus, Denmark
4. Kasper Jarnefelt, Architect, Helsinki
5. Ulla Broön, National Board of Social Welfare, Copenhagen
6. Eleonora Alaoui, Information Officer, The Swedish Institute, Stockholm
7. Eva Beverfelt, Norway Gerontology Institute, Oslo
8. Håkan Josefsson, Architect, White Architecture, Göteborg
9. Barry Korobkin, AIA, Boston, MA
10. Dean Painter, Eaton Terrace II, Lakewood, CO
11. Alan Black and Richard Webb of Rosewood Estate, Roseville, MN
12. Beth Sachs of Woodside Place, Oakmont, PA
13. Arvid Elness, AIA, Minneapolis, MN

There are a few people who affected my thinking and perceptions in important ways, including Paul and Terry Klaassen of **Sunrise** Retirement Communities of Oakton, Virginia. When Paul and I met we started our dialogue with a five-hour discussion of ideas, values, and future goals about housing and services for the frail. He invited me to speak at the Annual ALFAA conference in 1992 and took the risk of devoting a full day of the conference to a plenary discussion of this work. This allowed me to think about this book at a critical point in its development and select the best 500 slides from the 7000 I took. This review made the book a much better one in content and image.

Keren Brown Wilson, Ph.D., of Concepts in Community Care, Portland, Oregon, has been a consistently insightful source of ideas about care and management systems. Keren has taught me about the value of providing help tailored to the needs and lifestyles of people. This is a topic everyone discusses but few are able to deliver as well as she.

Jeroen Singleberg, Director of SEV, Rotterdam, the Netherlands, wrote back to me with enthusiasm, cleared his busy calendar to meet when I was there, and made most of the contacts for me in the Netherlands. Jacques Smit and his spouse, Rolie Post, befriended me and spent several days showing me work that reflected a range of issues in social housing experiments in Breda, the Netherlands.

Grete Bull hosted me in Oslo, identifying projects throughout Norway and providing me with evaluations of several buildings, including one I selected as a case study. Hans van Beek, of Pro Atelier in The Hague, the Netherlands, took me to several projects in Amsterdam and The Hague, and spent time describing his ideas about new directions in Dutch architecture. Jan Graafmans, Director of the Center for Biomedical and Health Care Technology at the Technical University in Eindhoven, the Netherlands, invited me to participate in an international conference on technology and aging. This gave me the opportunity to "debrief" at the end of my summer trip with a diverse collection of interesting people, including psychologists, economists, architects and product designers. It was a great way to start the note taking and storytelling that eventually resulted in this book.

Additionally, hundreds of people in the United States and Europe devoted as much as three hours each to interviews and facility tours. Their insights and commentary helped me to evaluate the strengths of each example.

LAST BUT NOT LEAST

The final group of people includes those who are the closest to me and have had the most influence on the work in substance and production. Far and above the most instrumental of these people has been Jenny Hamilton, a former graduate student, who word-processed and edited the entire manuscript, which forced her to learn how to read my writing (often mistaken for a cross between Korean and Sanskrit). She devoted tremendous energy to the project and was my single biggest fan and critic. This manuscript is much better as a result of her insights and questions. She mixed curiosity and perceptive commentary in a manner that made me see things in different ways and explain them better. Annette Wu, another talented former student, sandwiched time between taking the California architectural exam and preparing for graduate school at Harvard to help refine and redraw a number of the graphics.

A few others don't fit well into any of the previous categories but have influenced my work in important and different ways. Brian Hofland, Ph.D., Vice President of the Retirement Research Foundation, Chicago, listened to me and supported my work with ideas and money. He identified with the spirit of the project and advocated it at a number of levels. It is his term "fear-based system of regulation" that so aptly describes the situation we face in this country. Jerry Weisman, Ph.D., and Uriel Cohen, D.Arch, both of the Architecture and Urban Planning School at the University of Wisconsin, Milwaukee, provided very useful models for case study analysis and techniques for describing the transactional relationships of environment, care management, and social exchange through their work on dementia facilities. Jacqueline Leavitt, Ph.D., Associate Professor of Urban Planning at UCLA, Los Angeles, in the role of conference discussant, viewed the video lecture (Regnier, 1992) made of this work and provided insightful comments about what I had experienced and how it could best be communicated to designers and policymakers. She asked me to explain why older people are not present in many of the photos. This occurred because I always requested their permission before I took photographs and frequently found they were unwilling to be included. John Zeisel, Ph.D., of Building Diagnostics, Boston, is also a person whose fifteen-year association through many projects has given me insights, methods, and ideas. This field owes John an enormous debt of gratitude for his contribution to theory and communication.

Finally, I owe Judy Gonda, my spouse and best friend, the most credit for supporting me intellectually while serving as chief critic and mentor. I can't imagine how I could have left for three and a half months without her to shoulder the caregiving burden for our children, Jennifer and Heather. She, more than anyone else, made it possible for me to pursue these interests in a way that allowed very little to be compromised.

Efforts like this require a lot of support and I feel very lucky to be surrounded by such a generous and accommodating group of friends and supporters. Thank you, all!

PRECEDENTS, MODELS, AND PAST EXPERIENCE

1

INTRODUCTION

Using the term *assisted living* to describe the housing type that is the subject of this book, is easier in 1993 than it was in 1989 when we started exploring definitions of purpose-built housing arrangements for older frail people. Chapter Three frames a detailed working definition of assisted living by describing whom it serves and how to best provide the service.

Simply stated, assisted living is a long term care alternative which involves the delivery of professionally managed personal and health care services in a group setting that is residential in character and appearance in ways that optimize the physical and psychological independence of residents.

Its raison d'etre is to keep older adults out of an institution for as long as possible. Although the main component of assisted living involves personal care help with activities of daily living, the overall philosophy is one of personal self-management, seeking ways to allow the person or his or her family to manage a diverse range of health care services on an occasional or ongoing basis. As a housing type, it fits between congregate housing and skilled nursing care, but overlaps the traditional boundary of skilled care. As new forms of robotics and communications technology challenge the concept of institutional control, assisted living will become an even more popular avenue for caring for the frail.

Assisted living seems to have gained acceptance partly because it sounds more professional than "board and care," and it connotes independence more so than the term "personal care." However, the idea of caring for older frail people in residential-type environments can also be traced to the terms sheltered housing, residential care, assisted care, catered living, service flats, adult care homes, homes for the aged, and rest homes.

The term assisted living appears to have first been used by facility sponsors. Literature searches show that trade publications like *Provider*, *Contemporary Long-Term Care,* and *Retirement Housing Report* were the first to use it or to recognize its use in the retirement housing industry (Long-Term Care National Resource Center at UCLA/USC, 1989). Was the term popularized for advertising purposes, or is there a deeper meaning in its use? Keren Brown Wilson (1990) believes that in Oregon, where new rules were being promulgated for housing the frail, a fresh term that was different from the current federal and state lexicon was necessary to position professionally managed residential housing for the frail as a new idea. For many, the term connotes a philosophy about caregiving as well as an attitude about what constitutes an appropriate physical environment. Assisted living is separated from personal care, which has its origins in the nursing home industry (Regnier, Hamilton, and Yatabe, 1991). The licensing unit in personal care facilities is the individual bed, therefore it is not uncommon to hear institutional terminology like "patients and beds" used to describe both the setting and the people. Although assisted living is frequently regulated under the same laws as personal care, residential units and residents are the preferred reference.

The concept of assisted living has been described as a more professionally managed board and care setting, or a more intimate and residentially appearing personal care facility. Regardless of how exact the definition, assisted living has emerged as a housing concept for the frail that challenges our assumptions about how to serve an increasingly larger older age cohort.

FIGURE 1.1 *Sunrise* retirement community buildings are designed around the concept of a Victorian mansion house. Between forty-five to fifty units are accommodated within a three-story building, to achieve a minimum economy of scale for the production and delivery of health and personal care services to mentally and physically frail older people.

TWO TYPICAL RESIDENT PROFILES

Assisted living is often defined as a housing type for both the physically and mentally frail. In most facilities, however, these two groups are identified as having different needs to which the environment responds in different ways. In fact, debate and controversy continue over how compatible these two populations are and how they can best coexist. Assisted living facilities that isolate residents with major mental impairments are more common than those that "mainstream" populations.

Mentally Impaired Residents

Most segregated facilities isolate patients after they become a hazard to themselves or disturb others in a mixed context. These facilities often take care of dementia patients in smaller self-contained groups included within or adjacent to the main assisted living facility. Because memory impairments often increase with advanced age, they need to be treated like any other personal care need, rather than singled out as a reason for institutionalization.

The profile of the mentally impaired resident shows a physically active person experiencing disorientation, memory lapses, confusion, agitation, and frustration. Disorientation may involve confusion in a familiar setting, or failure to comprehend where his or her dwelling unit is located in the building. In some cases, symptoms are occasional or diurnal in their effects. In others, losses relate to short-term memory recall and can involve telling the same story or repeating the same comment several times in a conversation. As the dementing illness advances, restlessness, aimless wandering, occasional outbursts, and behavioral aggressiveness may develop. The line between acceptable and unacceptable behavior in an integrated assisted living environment may hinge on how the behavior of the demented individual affects other residents. When behavior becomes so disruptive that it increases the anxiety of other residents, isolation may be necessary.

Physically Impaired Residents

The second profile includes the mentally alert but physically frail older person who often suffers from one or more chronic ailments that restrict mobility, limit ability to carry out fine motor tasks, or cause problems with balance control. The probability is high that this person suffers from arthritis, hypertension, heart disease, diabetes, or hearing and visual impairments. It is also likely that assistance with bathing, dressing, medication supervision, toileting, ambulation, eating, or grooming will be needed, but these needs often stop short of twenty-four-hour supervised nursing care.

Assisted living can and does deal with both of these resident profiles. Neither resident profile should be relegated to a skilled nursing environment. Being disoriented and restless, just like being incontinent or disabled by arthritis, is not a good enough reason for institutionalization.

MEDICAL AND RESIDENTIAL MODELS OF HOUSING THE FRAIL

Recent studies of housing environments for Alzheimer's residents (Cohen and Weisman, 1991; Calkins, 1988) suggest specific furnishing and design adjustments that secure environments for the older mentally frail person. The research reinforces the idea that separate facilities are warranted, given the specific design mitigations which can be employed to help these people. The work, however, also implies that nursing home placement may be inevitable if a person lives long enough, and that facilities designed along a "medical" nursing home model are warranted. Nursing homes are currently the main institutional environment used to house the mentally frail.

Assuming that dementia residents must eventually be institutionalized foreshortens our vision and abandons the possibility of using the residential environment as the basis for therapeutic intervention. Swedish group homes like **Hasselknuten,** described in Chapter Five, keep demented residents in a homelike setting until death. In Sweden this approach is considered both feasible and appropriate. In the United States, noninstitutional housing and care models for dementia residents are in their infancy, but a few well-publicized models—the **Corinne Dolan Center,** in Heather Hill, Ohio, **Woodside Place,** in Oakmont, Pennsylvania, and the **Alzheimer's Center,** in Gardiner, Maine—have forced designers and administrators to rethink the problem and to respond differently than they did five years ago (Cohen and Day, 1991).

FIGURE 1.2 Personalization of single-occupied rooms was achieved at *Woodside Place,* in Oakmont,Pennsylvania, through a plate shelf with pegs mounted on two walls. This design was inspired by Shaker casework and helps to establish a friendly character.

As nursing home costs continue to escalate and public opinion of the nursing home as an acceptable alternative continues to wane, assisted living housing that increases control, choice, privacy, autonomy, and dignity will flourish. Europeans have recognized this trend. Their systems are not narrowed by regulation and shortsighted public policy. Our challenge in the United States is to pursue a housing policy for the frail that grants them the choice to stay at home when feasible and the option to live in a highly supportive group housing environment that preserves independence and enhances well-being.

The Medical Model

Nursing homes and assisted living are physically very different from one another. Assisted living has emerged as a type of housing rather than as an institution. The nursing home is modeled physically and operationally around the hospital. Its building codes specify wide doors to units and wide corridors for the exiting of patients in beds. Fire safety considerations narrow the range of acceptable wall, floor, and ceiling materials, and fire detection and monitoring equipment force many decisions that compromise appearance and privacy for the sake of safety.

Regulations that specify staff levels are tied to the number of beds served and the location of nurses' stations used to monitor those beds. Efficiency, driven by the desire to minimize required staff, dictates highly centralized plans. Distance maximums of ninety feet from the nurses' station to each resident's entry door require dense double-loaded corridor configurations with relatively narrow unit widths. This is why most "standard" two-bed rooms have beds placed side by side rather than toe to toe. The room shape, the semiprivate standard for privacy, and the distance from the room door to the nurses' station come directly from hospital experience. They have never been adequately tested for validity in a long-term care context.

The behavior of nursing home staff is also highly influenced by acute health care settings. Nurses and aides are trained to care for patients based on a medical model. This means that autonomy, mobility, and self-care are tolerated rather than encouraged, and that privacy, freedom, and dignity are subservient to efficiency and safety. In summary, the nursing home is a transformation of the acute-care hospital, which has never been carefully inspected for its appropriateness as a solution to the long-term care needs of a majority of the frail elderly.

The Residential Model

Assisted living, on the other hand, has grown out of the desire to keep older people independent in a comfortable residential setting for as long as possible. It emulates in character and style the board and care homes that are often remodeled hotels or large single-family homes. The precedent for the physical envelope of assisted living is the mansion house, country villa, or bed-and-breakfast hotel.

Assisted living has become the most common term in the United States for describing housing with services that fits between skilled nursing care and congregate housing. Although assisted living, as defined in this book, describes a *residential* alternative to nursing care, it has been used by different organizations and corporations in many different ways throughout the United States.

The Medical Versus the Residential Model

Many nursing homes concerned with serving the frail have remodeled conventional nursing home wards and called them assisted living. A lot of these projects are conversions of conventional two-bed nursing home rooms that have been carpeted and single-occupied. They conform to the rigid geometry of the nursing home, utilize wide doors and hallways, and contain a nurses' station just like institutional skilled care arrangements. Units are small, rarely have a kitchenette, and may require sharing a full bathroom.

Surprisingly, large corporations like Marriott Life Care and Forum, Inc., have also erred in the direction of creating assisted living units that are institutional in character. They have approached assisted living as a health care environment, placing it in the same building as the nursing home and referring to it as a "health center." In these models, assisted living is often located on a separate floor, but is stacked above or below skilled nursing.

The regulatory standards that establish the institutional geometry of the nursing home thus dictate the footprint of assisted living. The result is a narrow and deep dwelling unit with poor daylighting and inadequate square footage. The double-loaded institutional configuration and the appearance of a building constructed to nursing home standards impart an austere and unfriendly look from both the curb and the inside. Making assisted living conform to institutional standards is rationalized as adding to the project's economic flexibility. In this configuration, assisted living units can be easily remodeled into skilled care ones if future market forces dictate.

FIGURE 1.3 The porte cochere at the *Sunrise* retirement community in Fairfax, Virginia, is a secondary entrance. This facilitates pickup and drop-off activities while relieving the main entry of clutter from cars and pavement. The scale of the attachment is residential, further reinforcing the feeling of a large house.

Another bothersome aspect to this line of reasoning is the presumption that regulatory officials will find an institutional-appearing environment more suited to their preconceptions about quality of care. Some sponsors believe regulators feel more comfortable when a facility looks a bit institutional, because it fits their expectations of a health care facility. Touring multilevel facilities in the United States, one often finds nursing units and so-called assisted living comingled on the same floor, with the major differences being finishes and details. The most negative consequence of this comingled association of nursing care and assisted living is the message it sends to residents, visitors, and staff. It presents assisted living as an institutional health care environment, irreparably damaging the therapeutic philosophy of the setting. A few cosmetic adjustments to this "medical model" of assisted living cannot overcome its appearance as an institution.

Residential assisted living models recognize through their design that the size, scale, and configuration of a building creates the feeling of a homelike environment. In these models, residential housing forms the precedent rather than the hospital. In fact, many sponsors concerned with overly cautious regulatory agencies design their structures to meet institutional occupancy standards but then go to great effort specifying window, floor, and wall materials as well as finishes and casework details that have strong residential associations. Paul Klaassen, the developer of **Sunrise** Retirement Communities, argues that the codes, when carefully interpreted, are able to accommodate much greater flexibility that most presume.

It is the accumulated interpretations and the sheer volume of existing institutional solutions that cloud this situation, giving sponsors and regulators the perception that the codes are totally to blame. In fact, a major component of the problem may be their narrow, conventional interpretation. Unfortunately, each new assisted living unit that follows this narrowminded institutional model compromises the residential ideal and provides, through its presence, more evidence of how to incorrectly and uncreatively interpret code requirements.

In summary, the character, organization, and management philosophy of assisted living must reinforce its identity as a housing type and not an institutional building type. When assisted living is no more than a decorated nursing home, it loses its promise as a viable alternative to institutionalization. Under these conditions, management's desire to control efficiency and maximize convenience subverts the residents' need for privacy, autonomy, and independence.

PRECEDENTS AFFECTING THE RESIDENTIAL ASSISTED LIVING MODEL

Residential assisted living models have been influenced by a number of different precedents. Major influences include residentially styled personal care units in Continuing Care Retirement Communities (CCRC), northern European service houses, British sheltered housing schemes, sophisticated board and care facilities, and state-sponsored housing experiments. Each of these housing types has sought to reduce inappropriate placements of older people in nursing homes, and each has contributed to our understanding of how assisted living can help older frail people lead a more independent life.

1. Continuing Care Retirement Communities

CCRCs began as organizations focusing on the needs of older clergy and "deserving" older women. They have been present in the United States for about 100 years (Raper and Kalicki, 1988), but until the early 1960s they were considered a minor housing option for a small group of older people. Since that time their numbers have grown and currently over 700 facilities operate in the United States. Their special contribution to caring for the frail, is a product of the humane caring community created by the concept, the cooperation and mutual support it fosters among residents, and basic health care economics. CCRCs are designed to care for the lifetime nursing and health needs of residents, and are often structured to keep residents out of the nursing home component for as long as possible. When a CCRC is sponsored as a nonprofit corporation, savings associated with minimizing nursing home costs by keeping residents in assisted living can be used to offset fee increases and minimize first-time membership fees.

Furthermore, residents in these communities are important participants in management and activity programming. The community allows residents to grow old and frail with neighbors and friends, thus encouraging the frail to lead a more active and fulfilling life. The economic and social structure of the community encourages innovative ideas regarding therapy and treatment that keep residents as independent as possible.

CCRCs have been exploring different forms of assisted living for years. Both residential and medical models of assisted living exist in these communities. Many have identified assisted living as a key component in a concerted program to keep older frail residents physically and mentally independent. For example, **Mount San Antonio Gardens,** in Pomona, California, remodeled its assisted living building by adding physical therapy equipment and a part-time physical therapist. This change was made to involve residents in an exercise program designed to build their competency and forestall their move to the onsite nursing home.

Viewing assisted living as having a role in the rehabilitation, restoration, and maintenance of higher competence levels is an important philosophical and lifestyle attitude that can positively affect staff and residents. At **Mount San Antonio Gardens,** the residential appearance of the facility and its location on the edge of the campus away from the nursing center gives it an independent character when compared to medical models. Separating the facility administratively and financially from the auspices of the nursing home allows it to develop a philosophy of care that differs significantly from that of a health care environment.

2. Northern European and British Examples

Europe, which experienced an early demographic shift toward an aging society and progressive postwar social policies, initiated publicly approved housing programs for the elderly about fifteen to twenty years before the United States. Two case study books were published in the 1960s chronicling early experiences (Rutherford and Holst, 1963; and Beyer and Nierstrasz, 1967). More recent analyses have focused on trends and innovations (McRae, 1975; Goldenberg, 1982; Hoglund, 1985), special housing types (Rose and Bozeat, 1980, Heumann and Boldy, 1982), general planning guidelines (Weal and Weal, 1988; Valins, 1988), and research findings (Reinius, 1984; Pynoos and Liebig, forthcoming).

European policy choices have traditionally emphasized noninstitutional alternatives for the frail (ASVVO, 1991; Bull, 1987; Husbanken, 1987; Lindstrom, 1989). Early decisions reflected in building codes, experimental programs, attitudes toward independence, and financial commitments fueled progress in this area (Beckman, 1976). Older frail people, especially in Scandinavian countries, live in dramatically larger and more private residential environments. Widespread societal attitudes about "aging in place" and the absence of punitive regu-

FIGURE 1.4 Dwelling units in Scandinavian service houses are large in size and complete in their design. The kitchen in *Lesjatun,* in Lesja, Norway, is large enough to accommodate a table. The natural wood detailing is consistent with the rustic character of these dwellings located in a small rural town in the mountains.

lations that force patients to relocate as their health care needs increase have led to an enlightened attitude about caring for older people in a range of community contexts. Each European country has a unique housing stock based on its vernacular traditions and cultural differences (Daatland, 1991).

Geographical closeness and the tradition of visiting one another for study trips has led to many good ideas about housing for the frail, which are tested in several different governmental and organizational contexts. For example, the group home movement for dementia victims in Sweden has been copied in concept and interpreted by the Norwegians, Finns, and Dutch in their own particular ways. Often they make major improvements or test out significant alterations when adopting an idea.

European Design Competitions

Another interesting aspect of the European system is its reliance on architectural competitions for testing new ideas about architecture, service planning, and caregiving. A good example of this is the Swedish competition organized by SPRI (Swedish Planning and Rationalization Institute) in the early 1980s to develop new prototypes for what were called "local nursing homes." Unfortunately, the results of this competition were never fully realized because of changes in housing policies that discouraged the building of new nursing homes. But, eight innovative solutions were published based on the concept of decentralizing the nursing home into sixteen-bed clusters (SPRI, 1980). Courtyards, outdoor patios, living room lounges, dining rooms, therapy areas, and activity spaces were used to create smaller-scale, homelike unit clusters. These designs were more residential in character and better integrated with the landscape environment than previous hospital-like models. Recent SPRI research has focused on identifying exemplary group homes (SPRI, 1990) and strategies for remodeling older nursing homes that increase privacy and autonomy (SPRI, 1989).

Another influential competition in Finland (Finnish Architectural Review, 1987) explored ideas about how to combine community health services with an "old peoples home." The winning entry that opened in the spring of 1991 in Oitti, Finland, clustered twelve units into several small "villages." Common rooms with tall pyramid-shaped skylights introduce natural light along a fragmented spine linking four village "clusters." Community health services and more intensive in-patient care are located at one end of the linear building, with residential clusters at the opposite end.

A third recent competition in the Netherlands explored new ideas for elderly living within the city (Stichting Wonen Amsterdam, 1987). Twelve young architects were given the opportunity to work with social housing on four urban sites in Amsterdam. The solutions reveal a range of ideas about how architecture can facilitate new ways for older people to engage directly and vicariously with the surrounding city context. The work ranges from artful and expressive ideas with awkward social implications to solutions based on social principles that are meant to reduce dependency while increasing autonomy, privacy, and a sense of community. In all cases, the work explores new ideas with boldness and curiosity.

In Europe, social service, health care, and housing agencies recognize the influence of new architectural forms in challenging conventional thinking about social problems. This is a lesson American government can learn from the Europeans. Architecture is an instrument of change that can stimulate new ways of responding to problems.

Neighborhood-Based Approach

Probably the most extraordinary aspect of European housing stock is how it adds to the social welfare of the surrounding community by sharing common services. Swimming pools, restaurants, health services, physical therapy equipment, recreational programs, and meeting room space are designed to be used by residents and older people living in the surrounding neighborhood. The general assumption that facilities for the aged are for all older people in the community gives housing with services an important place in the community. It combines the social potential of the senior center with the health security of a clinic and the personal autonomy of independent housing. It is a powerful combination with both symbolic and substantive meaning to community members. Curiously, the United States system seems to have encouraged separation through government agencies that rarely co-venture and zoning policies that discourage mixed land uses. Our system lacks the synergism and identity that concepts like co-location have generated in European models.

3. Experimental Assisted Living Projects

In the United States, experimental housing arrangements have been conceived to test new ideas. Many of these have received support from professional groups and have been subject to thorough evaluations. What makes most of these examples powerful is that they combine building design with ideas about caregiving.

Captain Eldridge Congregate House

One influential building in this regard is the **Captain Eldridge Congregate House,** in Hyannis, Massachusetts, which combines new construction with adaptive reuse (Morton, 1981). This building was one of approximately twenty congregate houses created by the Massachusetts Executive Office of Communities and Development and the Department of Elder Affairs in the late 1970s and early eighties. It embraces a "family" lifestyle and is targeted toward at-risk seniors living in the community or inappropriately placed in nursing homes. Evaluation and commentary regarding the success of this building and other congregate house examples are contained in the book *Independence Through Interdependence* (Welch, Parker, and Zeisel, 1984).

The **Captain Eldridge Congregate House** appears later as a case study (Chapter Five). It was designed by Barry Korobkin, who earlier had worked in the office of the Dutch architect Herman Hertzberger. John Zeisel, the design behavior consultant, authored behavioral design hypotheses that were used to make critical design review adjustments to the plan and to create a foundation for evaluation. The scheme was copied by dozens of other projects throughout Massachusetts and became an important and influential precedent for later projects like **Elder Homestead** in the Minneapolis suburb of Minnetonka. **The Captain Eldridge Congregate House** was classified as a boarding home and thus was designed without one-hour fire separations between the resident units and hallways. A small porchlike alcove, in combination with Dutch doors, a double-hung window, and a "porch" lamp, connected each resident's unit to an enclosed two-story light-filled central atrium. The solution created a sense of whimsy and delight for residents, giving each unit a separate identity and improving on Hertzberger's original entry idea in **de Drie Hoven** project in Amsterdam (Altenheim in Amsterdam, 1976; de Drie Hoven Old People's Center, 1976; and Old People in Amsterdam, 1977). What makes the **Captain Eldridge Congregate House** unique and convincing is the scale of the building and its stylistic fit into a neighborhood of older two-story Cape Cod homes.

Annie Maxim House

Korobkin went on to design another congregate dwelling, the **Annie Maxim House,** in Rochester, Massachusetts, which continued the idea of the porch entry linking the unit to common spaces through a one-story horseshoe-shaped, single-loaded balcony corridor (Boles, 1985). This plan ensures privacy by replacing the atrium with a landscaped courtyard that overlooks a nearby lake. The common spaces are generous and vary in character. Interior double-hung windows throughout the building give various common spaces a friendly, transparent appearance which facilitates previewing.

Elder Homestead

This project, in Minnetonka, Minnesota, borrowed the Dutch door and double-hung window features from the **Captain Eldridge Congregate House**, while resolving code compliance difficulties by defining each four-unit cluster as a shared dwelling (Gaskie, 1988). Fire separations were placed between the exit corridor and a "cluster parlor." The imagery of this building took its inspiration from the vernacular of the rural Minnesota farmhouse. Porches, sloping roofs, dormers, and a relatively complex plan created by the clustering of units give it a rambling, friendly look reminiscent of the vernacular rural housing stock of the region. Interior details like freestanding wood columns, antique furnishings donated by residents, and finishes based on residential standards continue this friendly, informal style on the inside.

FIGURE 1.5 The *Annie Maxim House* in Rochester, Massachusetts, has a horseshoe-shaped plan that creates a central courtyard. An exterior porch and single-loaded interior breezeway link dwelling units to a common building where food preparation, social activities, and the laundry are located. Units are small, but each has a kitchen and a bedroom.

Regency Park and Park Place

An oversupply of congregate housing in Oregon in the early eighties encouraged the development of an experiment through the 2176 Medicaid Waiver program. Keren Brown Wilson, working with two Portland housing projects, **Regency Park** and **Park Place,** moved patients from nearby nursing homes into these relatively conventional congregate housing facilities. A resident-assessment program and careful monitoring of needs and special problems led to a successful program that helped very old and frail residents maintain independence. This experiment called into question assumptions about the necessity of institutionalizing older frail people.

The program also saved considerable money by substituting a less expensive assisted living unit for a nursing home bed. The success of this program and its ability to deliver a higher standard of housing and more personalized care, for less money, has encouraged other experiments. The regulations tested by Oregon for the program have been made into law and are being adopted by other states. Wilson has gone on to develop several other smaller-scale projects (**Rackleff House** and **Juniper House**) in small towns adjoining the Portland metropolitan area. The dwelling units in these projects are studio and one-bedroom designs complete with a kitchen and full bathroom.

Alzheimer's Care Center

Other interesting experiments have focused on residentially based facilities for dementia victims. One influential program, which embraced a group home philosophy, was the **Alzheimer's Care Center**, in Gardiner, Maine. Opened in 1988, its primary contributions have been in the creative application of residential imagery to the building's design and the development of a highly individualized care program for residents. Other innovative features include an interior wandering pathway that loops through the building, clustering its seventeen units into three smaller groups (Cohen and Day, 1991).

Corinne Dolan Center

This Alzheimer's facility in Heather Hill, Ohio, has gone a step further by organizing a research program to test aspects of the building's design. Two identical triangular mirror-image wings, modeled after the successful Osmund plan of the Weiss Institute at the Philadelphia Geriatric Center (Liebowitz, Lawton, and Waldman, 1979), are joined together where the building's entry and administrative services are located (Cohen and Day, 1991; Calkins, 1988). An early choice not to license the building as a nursing home allowed more experimentation with architectural aspects. Support from the Robert Wood Johnson Foundation has allowed the two identical wings to sustain a range of controlled intervention studies. A number of experimental ideas use the environment to overcome memory deficits. These include unusual and

FIGURE 1.6 *Elder Homestead* in Minnetonka, Minnesota, was inspired by the imagery of the rural farmhouse. The asymmetrical form, steep roof pitch, attached entry porch, three-season screened porch, and residential detailing were taken from rural precedents as a source for organization and character development.

controversial ideas such as designing interior windows that allow toilet fixtures to be seen from the bedroom in an effort to reduce incidents of incontinence as well as testing the effectiveness of ideas such as personalized showcase boxes at the unit entry, to remind residents which unit is theirs.

Woodside Place

A most recent and promising new residential design for the mentally frail is **Woodside Place,** in Oakmont, Pennsylvania (*Contemporary Long-Term Care,* 1992a, 1992b). This project, described in more detail in Chapter Five, has several noteworthy features. The design concept involves the use of three small "houses," of twelve residents each, linked to one another and to common services through an enclosed wandering pathway. Thus the intimacy of a small group is achieved, while economies of scale are sufficient enough to support a self-contained kitchen and a variety of common spaces for a range of social and therapeutic activities. Designed to look like a Shaker commune from the outside, the juxtaposition of outdoor areas for views and use and the creation of a friendly, comfortable interior that accommodates wandering and informal socializing are among the project's most compelling features.

Rosewood Estate

One of the most controversial explorations in providing care to the mentally and physically frail has been through housing with meal services and medical and personal care services coordinated through home care agencies situated in the building. **Rosewood Estate,** in Roseville, Minnesota, is one such well-known example, and appears in more detail in Chapter Five. Most residents who move there choose between this setting and a nursing home. Designed by Arvid Elness, AIA, after his experience with **Elder Homestead,** in Minnetonka, Minnesota, it contains sixty-eight units in a three-story building that overlooks a large lake. From the street it appears as three attached houses. The colonial-style architecture fits into the surrounding suburban residential fabric. The dwelling units are large and are located around eighteen decentralized lounges. Common rooms for activities, meals, and services are located near the center of the building. An agreement with a nearby nonprofit home health care agency allows residents to purchase health and personal care services on demand or by advance reservation, in fifteen-minute increments. Purchasing services on an "as needed" basis allows the family to participate in caregiving and care management. Service billing is handled separately, with total monthly costs averaging 20–25 percent less than for a private skilled nursing home.

This model, although considered illegal in many states because it operates without a license, allows increased flexibility. It is similar in concept to the Danish and Swedish models of housing and service provision. The fact that regulations keep this model from being implemented in many states is further evidence of the impediments facing sponsors who seek to develop new housing alternatives for the frail.

Each of these experiments has advanced the state of the art by challenging existing regulations, by utilizing homelike residential designs to encourage social exchange and friendship formation, and by reducing project size to the minimum for efficient operation. The lessons learned from these experimental models have shaped the options available today.

4. Sophisticated Board and Care Arrangements

Board and care housing is the most common noninstitutional alternative for older frail people. In 1988, there were reported to be nearly 600,000 beds licensed for the elderly and disabled in approximately 40,000 facilities throughout the United States (US Senate Select Committee on Aging, 1989). However, the same report estimated that an additional 400,000 people could very well be living in unlicensed arrangements in the country. Board and care beds now rival nursing home beds in number. Between 1984 and 1988 there was a 71 percent increase in the number of licensed beds.

Approximately half the residents have come from mental institutions in states where economic pressures have forced state governments to relocate them in less costly board and care settings. Nursing homes or mental institutions can be 3.5 to 5.5 times more costly, respectively. Nearly 75 percent of these settings rely on SSI as the only source of support, which means that residents pay approximately $600 per month for room and board. Concerns about how some residents are mistreated have prompted a GAO report and a Senate subcommittee investigation (U.S. Senate Select Committee on Aging, 1989). These reports are focused unfortunately on those who abuse the system (11,000 complaints were received by state licensing offices in 1987).

Small Board and Care Settings:

The positive attributes of facilities doing a good job at reimbursement rates far below those charged by institutions have not been the focus of much research. The vast majority of these settings are small family-type arrangements of less than seven residents. Turnover rates are substantial and can be a major problem. One Oregon study of the adult foster care system reported 20 percent of licensed providers left the industry in one year, 1989 (Kane, Illuston, Kane, and Nyman, 1990).

The idea of taking care of older frail people in small-scale settings, where individual attention and care are provided as a philosophical extension of family care, seems very appealing when compared with treatment in the standard nursing home. In reality, some of the best and worst situations are encountered in these arrangements. Their small size makes them intimate, and residential in character, as well as very difficult to monitor. In contrast to nursing homes, they are highly unsophisti-

FIGURE 1.7 The Swedish small group home is a professional model of board and care housing for six to eight people. The *Strand,* in Arvika, Sweden, uses a simple pitched ceiling spatial extrusion to create a large comfortable space for living and dining activities. The upper floor of this two-story dwelling is offered to families, while the ground floor is for mentally and physically frail older people. The laundry, located in a neutral first-floor location, is shared with families.

cated, but their setting is also less overwhelming and less institutional. Because these facilities don't provide nursing services, board and care providers may be more committed to keeping the older frail out of nursing homes.

Board and care arrangements are characterized as being managed by people with big hearts but little formal training. However, the constant turnover in this housing sector and dishonest sponsors who violate public trust have tarnished its reputation. Physical environments are often not purpose-built and contain a majority of double-occupancy rooms. Privacy and autonomy are difficult to achieve, especially for homes that cater primarily to low-income clients. Regulatory differences between states and the difficulty in overseeing so many small facilities make the potential for abuse a major problem. One of the main policy concerns facing the board and care industry centers around the regulation and classification of these arrangements in various states (Benjamin and Newcomer, 1986). AARP in a recent publication has suggested model guidelines for licensure in an effort to establish continuity between states (Dobkin, 1989). A recent evaluation of foster care and assisted living in Oregon (Kane, Illuston, Kane, and Nyman, 1990) juxtaposed cost and care advantages with the difficulty in achieving predictable quality control. This is a dilemma that is not likely to be resolved soon.

Sunrise Retirement Communities

One of the most interesting corporate examples of a new breed of sophisticated board and care housing is the **Sunrise** Retirement Community in Oakton, Virginia. The **Sunrise** community in Arlington, Virginia, is profiled in Chapter Five. The Sunrise corporation began as a small-scale family operation in 1980. During the last twelve years it has advanced to become one of the fastest-growing small businesses in America, employing over 700 people in 1992. The strategy of Paul and Terry Klaassen in achieving this success was to focus on the provision of high-quality personal care in new buildings, large enough to operate efficiently but small enough to resemble a large single-family house or a bed-and-breakfast hotel. They chose a Victorian style and copied residential design details from older mansion houses.

The most noteworthy aspect of their success, however, has been their ability to recruit, train, and motivate caregiving professionals. They are dedicated to testing new ideas and have never built the same project without making significant alterations. Each project embodies a similar basic concept but supports experimental idea testing. The model has been carefully developed through a heuristic approach that is evaluated and refined on the basis of feedback from residents, family members, and staff. Residents are typically upper-middle-income private payees, the majority of whom have family members nearby. The Sunrise approach could be characterized as "hands on" care provision in small but private units, in residentially styled buildings located in attractive neighborhoods. It is surprising, given the need for this type of housing and service combination and the success of Sunrise, that more board and care providers have not emulated their approach.

SUMMARY

Although numerous influences from foreign and domestic models have stimulated the thinking behind today's assisted living prototypes, there is still a great deal of disagreement over where this housing and service combination is headed in the future. Most trade journal articles dealing with assisted living have been more editorial than empirical. One questionnaire by *Contemporary Long-Term Care* sought responses from 200 operators of assisted living facilities. A series of articles (Seip, 1989a; 1989b; 1989c; 1990) reflected the wide range of providers and their services. Only a third to 40 percent appeared to approach model definitions of assisted living as set forth in Chapter Three. For example, 33 percent had none or only some units carpeted and 30 percent had none or only some units with a private bathrooms. This indicates as many as a third of the facilities surveyed have physical design attributes we would normally associate with institutions. On the other hand, 22 percent did not include assistance with dressing or bathing and 41 percent did not include toileting assistance in their monthly fee. These facilities appear to be less service-intensive than most assisted living arrangements. Kalymun (1990) reviewed ten facilities focusing on unit plans, available services, and the management's characterization of an expected resident profile. The most interesting outcome of this research was the range of variability in models. Clearly, more specific definitions and criteria for quality and excellence are needed to guide the future development of assisted living.

THE PURPOSE AND PROBLEMS OF HOUSING FOR THE FRAIL

2

INTRODUCTION

Two fundamental beliefs underlie much of the work appearing in this book.

First: Many older frail people currently living in nursing homes can lead a better, more satisfying life in a less institutional environment.

Second: Group housing alternatives that are residential in nature, invite family participation, and are based on a therapeutic model of caregiving can and should replace many nursing homes.

Older people should stay in their own home as long as they want to, with the aid and support of family members and home health care personnel. Once their needs outpace the ability of home care to serve them and the capacity of the family to provide informal support, they should be able to move to a group living arrangement where they or their family can maintain as much control as possible over their life. This setting should be designed to foster emotional support, social engagement, intellectual stimulation, and independence within a residential context. Furthermore, it should contain private dwelling units with access to physical and occupational therapy.

Although this position may appear a touch idealistic, our society has undergone numerous changes that favor highly supportive residential housing in place of institutional settings. Further support for this assertion can be found in the following trends.

1. The increasing numbers of people over the age of eighty-five,
2. The increasing costs of long-term care,
3. The increasing popularity of home care and community care options,
4. The increasing resistance of older people to accept institutionalization,
5. The increasing degree to which new communications and robotics technology can keep older people independent,
6. The increasing concern on the part of state government to reduce costs by discouraging institutionalization,
7. The increasing consumer demand for flexible arrangements that maximize choice, and
8. The increasing interest on the part of corporate providers, builders, and nonprofit sponsors to satisfy consumer demands for highly supportive, but noninstitutional housing options.

These eight factors have dampened interest in skilled nursing care, while encouraging the development of residential alternatives to institutionalization.

FOURTEEN QUESTIONS

The remainder of this chapter is organized around fourteen questions that seek to clarify the purpose and meaning of assisted living. They focus on the role the physical environment plays in supporting and enhancing assisted living housing.

1. What are the basic qualities and characteristics of the assisted living housing type?
2. What is the current and future demand for assisted living housing?
3. What broader roles can assisted living play in the community?
4. How and what can we learn from northern European housing models for the frail?
5. What are the costs and how can they be managed?
6. What contribution does architecture make in furthering the therapeutic goals of assisted living?
7. What role can the family play in helping the older frail person?
8. How does regulation constrain innovation, experimentation, and progress?
9. What elements constitute a therapeutic environment?
10. Is institutionalization inevitable?
11. How can families and institutions work together to support the needs of the older frail person?
12. How can the environment enhance independence and privacy?
13. Should the mentally and physically frail be integrated or segregated?
14. How can a compassionate and competent staff be recruited, trained, and retained?

The answers to these questions reveal the potential of assisted living settings to support the needs of the older frail in noninstitutional residential environments.

1. WHAT ARE THE BASIC QUALITIES AND CHARACTERISTICS OF THE ASSISTED LIVING HOUSING TYPE ?

Introduction

The basic idea behind assisted living is to provide professionally based personal and limited health care to vulnerable frail people within a residential rather than institutional environment. It can be viewed as combining the residential qualities and the friendly scale of board and care housing with the professionalism and sophistication of a typical personal care setting targeted toward residents who in the past would have normally resided in intermediate and skilled nursing facilities.

Because industry definitions are vague and often distorted by marketing descriptions that rarely clarify basic characteristics, the following nine definitional qualities are suggested as a loose normative definition. A more detailed treatment of these defining characteristics is presented in Chapter Three. Although few assisted living facilities can meet all of these definitional qualities, these criteria nonetheless provide appropriate targets for the development of highly supportive, humane residential housing for the mentally and physically frail.

Definitional Qualities

1. **Appear Residential in Character:** The form and character of assisted living should be derived from the house and not the hospital.

2. **Perceived as Small in Size:** The setting should be as small as it can be without sacrificing monthly cost stability and the capability to provide twenty-four-hour assistance.

3. **Provide Residential Privacy and Completeness:** The housing unit should be complete, with a full bathroom and at least a kitchenette.

4. **Recognize the Uniqueness of Each Resident:** Assessment and treatment plans should be customized to a resident's abilities, disabilities, and interests.

5. **Foster Independence, Interdependence, and Individuality:** The focus of care should be on self-maintenance with assistance. Residents should help themselves and one another.

6. **Focus on Health Maintenance, Physical Movement, and Mental Stimulation:** Services should be aimed at stabilizing decline, improving competency, and building reserve capacity.

7. **Support Family Involvement:** A caregiving partnership between the facility and the family should be forged that shares responsibility for resident well-being.

8. **Maintain Connections with the Surrounding Community:** The setting should integrate rather than isolate residents from community resources and contacts.

9. **Serve the Frail:** Residents should be older people in danger of institutionalization because of their needs for assistance and support.

2. WHAT IS THE CURRENT AND FUTURE DEMAND FOR ASSISTED LIVING HOUSING?

Introduction

The single biggest factor affecting the demand for assisted living housing is the growth of the oldest-old population. Middle series projections by the United States Bureau of the Census anticipate a doubling of the 85+ population in the twenty years between 1990 and 2010 followed by a further doubling in the next thirty years (2010–40). This 300 percent predicted increase in fifty years is considerable, but may greatly underestimate the potential increase. Guralnik, Yanagishita, and Schneider (1988) argue that pending biomedical breakthroughs and advances in disease prevention and therapy could continue the 2 percent decline in mortality experienced by the United States in the last twenty years. Using this assumption, their forecast shows the 85+ population spiraling from 3.3 million in 1990 to 23.3 million in 2040. This represents an over 600 percent increase in this vulnerable age group.

Table 2.1 ALTERNATIVE PROJECTIONS OF LIFE EXPECTANCY FOR THE POPULATION 65+ AND 85+; 1980, 1990, AND 2040

	1980[1]	1990[2]	High Mortality 2040[3]	Middle Series Mortality 2040[3]	Low Mortality 2040[3]	Assumption of 2% Annual Mortality Decline 2040[4]
Life Expectancy from birth (yrs)						
Male	70.0	72.1	72.7	75.0	77.8	85.9
Female	77.4	79.0	80.3	83.1	86.7	91.5
Population (thousands)						
Age 85+	2,240	3,254	9,648	12,834	17,258	23,319
Age 65+	25,549	31,559	60,227	66,989	75,159	86,805
Percentage of 65+ Population Age 85 and Over	8.7%	10.3%	16.0%	19.2%	23.0%	26.9%

1. U.S. Bureau of the Census, tabulated from Decennial Census of Population, 1983.
2. U.S. Bureau of the Census, 1989.
3. U.S. Bureau of the Census, 1984.
4. Guralnik, Yanagishita, and Schneider, 1988.

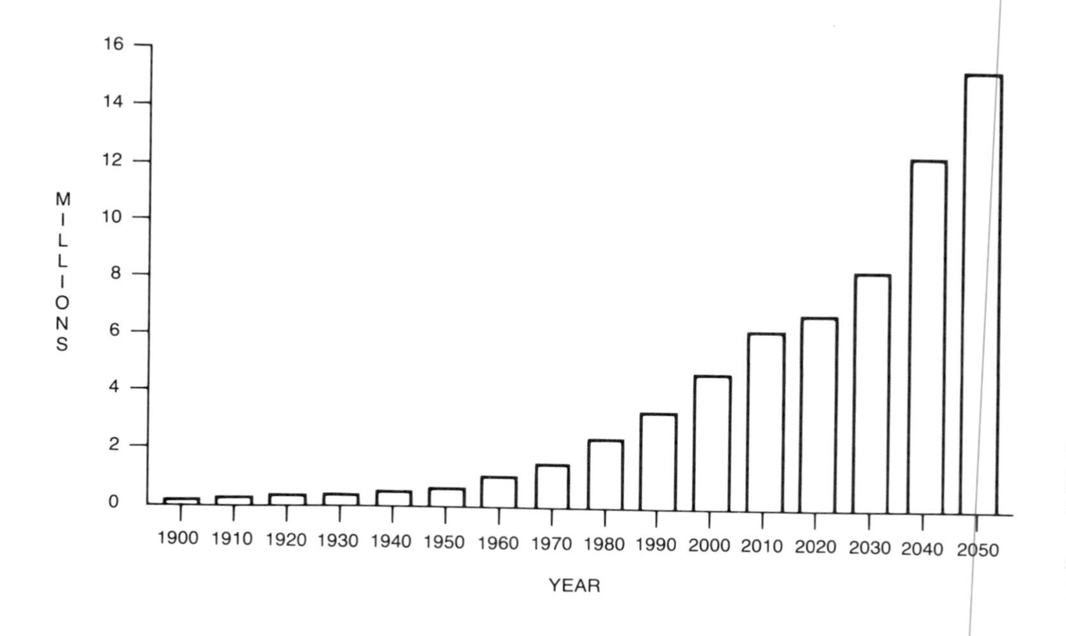

CHART 2.1 The actual and projected increase in population for 85 years and older from 1900 to 2050 (Source: US Senate Select Committee on Aging, 1991).

International Trends

The trend toward an older, frailer population is not just an American phenomenon. Japan is projected to experience a 274 percent increase in the 80+ population between 1990 and 2025 (U.S. Senate Special Committee on Aging, 1991). However, the major numerical increases in the oldest-old world population growth will be in less-developed countries. They will outpace the growth of more developed countries in the next thirty-five years by a factor of 3 (Myers, 1990). Today, the problems of the oldest-old rest with North America, Europe, and Japan; but by 2025, China (with 25.2 million oldest-old) and India (16.4 million) will surpass America's lead by pushing the United States from first place in the world's population of 80+ people into third (2025—14.3 million) (U.S. Senate Special Committee on Aging, 1988).

The Older-Old and Service Need

The older the population, the greater the likelihood that formal service intervention will be necessary to support a person's independence. Rivlin and Weiner (1988) estimate more than 60 percent of the current 85+ population is disabled as compared to 18.4 percent of the younger-old (65–84). Although informal care provision accounts for the vast majority of help provided to the elderly, as the population ages, the percentage that relies on formal support increases (Chappell, 1990). For example, while only 1.1 percent of the younger elderly (65–84) are in nursing homes, this percentage increases twenty-fold to 22.2 percent for the 85+ age group. Of those over the age of 65 who are severely disabled,[1] approximately 35 percent are in nursing homes. Thus the probability that more intensive assistance from the formal support system will be needed grows dramatically with an increase in age.

In 1988, approximately 6.9 million people over the age of 65 were in need of long-term care. At the turn of the century, this will increase 30 percent to 9 million. During the following forty years, it will double to 18 million (U.S. Senate Special Committee on Aging, 1991). Current nursing home stock will only care for a fraction of this demand. Home care and alternative noninstitutional housing and service arrangements will be the answer for the vast majority of this continually growing segment of the older frail.

[1] Severely disabled is defined as needing assistance with three of the six activities of daily living—eating, toileting, bathing, dressing, getting in and out of bed, and getting around inside (Rivlin and Weiner, 1988).

3. WHAT BROADER ROLES CAN ASSISTED LIVING PLAY IN THE COMMUNITY?

Introduction

One of the tragic consequences of current United States policy is the way housing for the frail is isolated from the surrounding community. Institutions, by history and habit, tend to be separated from other community functions. To overcome their isolation, they must strive to connect, link, and integrate their activities with those of the surrounding community. Universities are a prime example. The expression "town and gown" describes the estranged relationship between the university and the community, which can at times be both tenuous or transactional in nature.

Land Use Segregation

In the United States, housing for the frail is segregated by zoning criteria that reflect this bias. Land-use considerations relevant to housing, such as nearby services, reduced noise, increased public safety, neighborhood parks, and accessible retail establishments, are rarely thought necessary for nursing homes. Housing for the older frail can also require institutional zoning, which has few benefits and some clear disadvantages.

Europeans, on the other hand, have developed systems of social and community connection by overlapping retail, service, and residential land uses. Housing for the frail is often mixed with other land uses. These often combine the community aspects of a senior citizens center, the residential attributes of housing for the frail, and the service support capability of a home care agency or community health center. These settings offer social and health services to older people in the surrounding community as well as those who live in attached housing.

Community Service Delivery

Two strategies are present in the conceptualization of European service production and delivery. The first involves manufacturing or creating services at the site, which are then delivered to residents in the surrounding neighborhood. Common services produced in this way include meals-on-wheels, home help, and home health care. These activities are organized within service houses located in neighborhoods throughout the city. Emergency care is also organized this way. Call buttons from attached residential units are combined with telephone hookups from people living in the neighborhood. They are monitored by the service house where help is dispatched.

FIGURE 2.1 Eighty of the 100 site-visited European projects contain restaurants open to both project residents and older people living in the surrounding neighborhood. These attract people to the site for social programs, giving the place the feeling of a community center rather than an institution.

Multi-Purpose Single Site

The second strategy involves visiting the service house to utilize health care, nutrition, physical therapy, occupational therapy, and recreational services. In this model, neighborhood residents, as well as residents of the service house, receive help through community programs. Programs like adult day care, respite care, physical therapy, and occupational therapy are available. In some buildings, major investments in common spaces such as swimming pools, rehabilitative health equipment, and gymnasium spaces are financed through the participation of municipal entities that co-own these facilities, making them accessible to a range of community groups. This form of public investment changes the perception of the place in the minds of community residents. The service house is a place to seek help with problems, to enjoy fellowship and recreation, and to assist older residents to stay independent in the community. The co-location in the service house of children's day-care centers and restaurants open to all age groups provides reasons to regard it as a community resource and not just a site for older sick people.

4. HOW AND WHAT CAN WE LEARN FROM NORTHERN EUROPEAN HOUSING MODELS FOR THE FRAIL?

Introduction

One intriguing attribute of assisted living housing in northern European countries is the unique diversity of solutions available to meet geographic and cultural circumstances. Although Sweden, Norway, Denmark, Finland, and the Netherlands have used the service house as the basic building block for providing housing and services to the frail elderly, solutions reflect an amazing amount of local diversity.

FIGURE 2.2 Urban mixed-use housing projects are common in Finland. This one contains a pharmacy, health clinic, and child-care center. The children's playground is located in an L-shaped courtyard, protected from the noise and danger of the street by the housing block. Residents have balconies that overlook this courtyard and can watch children play.

Similarity with Diversity

In Denmark, physical and occupational therapy (ergo therapy) are part of the daily life in the setting. In Sweden, experimentation with small group homes for dementia victims has resulted in a new model of care. These buildings appear very residential in character and through their design they encourage residents to engage in activities of daily living, an approach that reinforces the resident's sense of independence. In Finland, sponsors of mixed-use buildings have refined ideas from Sweden and Denmark, with sophisticated and urbanistic architecture. In the Netherlands, many experiments center on how to mainstream nursing home patients with ambulatory residents in settings that encourage independence. Finally, in Norway, projects in rural and small-town contexts demonstrate a sensitivity to the scale of the local community and the building materials and traditions native to the region.

Another major difference between the United States and the northern European systems of care is the financial commitment northern Europeans have made to providing high-quality housing and services for middle- to lower-income populations. Funding comes from a combination of national health care insurance and housing subsidies. The housing standard for the frail elderly is higher than comparable housing in the United States. In Europe the strategy is to deliver health and personal care services to purpose built housing for the frail as well as older people living in normal housing in the community. Housing is conceptualized as a "service" to older frail people.

European System Attributes

Another noteworthy aspect of the European approach is its commitment to keeping older people independent for as long as possible. This commitment starts with a broadly based home care system and is backed up by service housing for those with the most intense needs (Soderstrom and Viklund, 1986). Regulations have not circumvented this commitment. Only severely impaired dementia victims with behavioral problems and complicated medical cases are considered to need institutionalization.

The degree of experimentation with new housing and service combinations is especially advanced in the Netherlands (WVC, 1989). Public agencies and nonprofit sponsors are continually devising new project experiments that test ideas about family care, case management, and self-maintaining philosophies over five- to seven-year evaluation periods. Professionalism in the Netherlands appears solid, reflecting the society's positive assessment of formal care providers and their central role in maintaining and nurturing the sick and weak. Care professionals are given greater autonomy and rarely find individual initiative stifled by regulations that confuse their commitment to helping older people.

One of the few less-advanced aspects of European service systems involves examples of well-organized family partnerships within housing projects. Even though these countries encourage family participation through social interaction, care sharing partnerships are relatively rare. Family members view the provision of professional help for the aged as a government entitlement. Many feel they have contributed taxes over a lifetime to finance the system and therefore care is increasingly seen as a government rather than family obligation. Furthermore, the high level of female employment makes it difficult for working daughters to accept extended responsibility for aged parents. However, the broad range of community-care supports, like adult day care, respite care, and home care, make it possible for families to more easily manage the older frail at home.

Northern European sophistication in systems of health and social care considerably outshines that of United States efforts. Mistakes have been made as these European systems have developed, but on balance they continue to offer the most creative and comprehensive models of care in the world.

5. WHAT ARE THE COSTS AND HOW CAN THEY BE MANAGED?

Introduction

In the United States, the cost of health care and long-term care for the elderly, as a proportion of all federal spending on the elderly, has increased from 6 percent in 1960 to an estimated 32 percent in 1991 (U.S. Senate Special Committee on Aging, 1991). If this trend continues, in fifty years health care costs will equal the average person's social security entitlement. A big factor in these increases has been the cost of long-term care. Currently, $33 billion are devoted to paying the yearly cost of nursing home care, but by the years 2016–20 it is predicted that $98 billion annually (1987 dollars) will be needed (Rivlin and Weiner, 1988). Currently, about half of the average $22,000 per year cost of nursing home stays is paid out of pocket by older people. The other half is paid primarily by Medicaid and Medicare (Rivlin and Weiner, 1988).

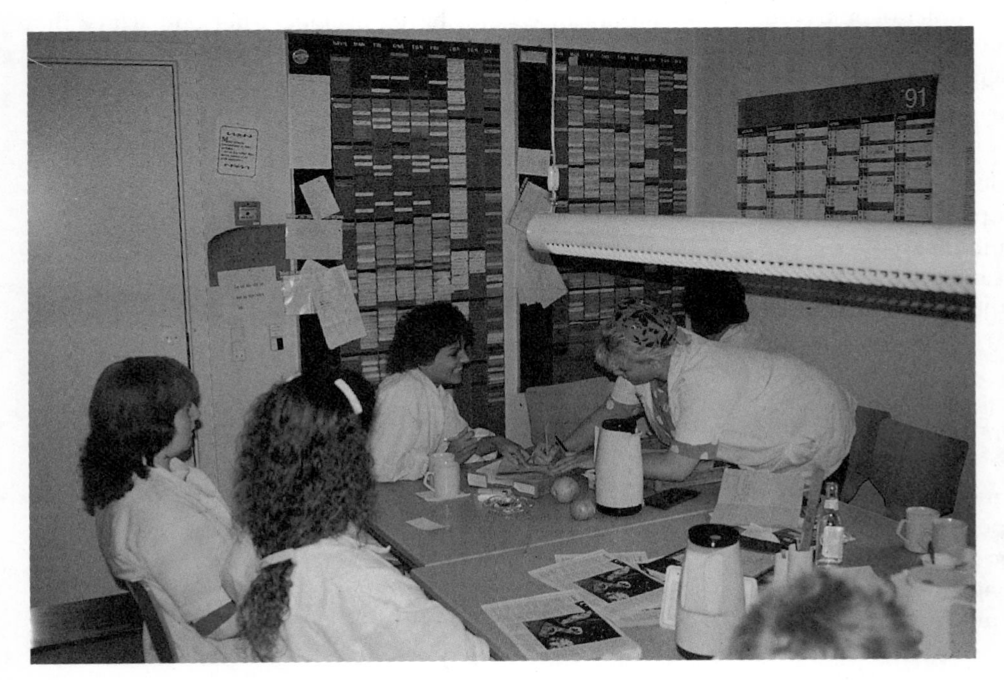

FIGURE 2.3 European service houses typically contain meeting space for home health agencies that provide personal and health care services to older people in the surrounding neighborhood, as well as to project residents. With service houses organized this way, health care professionals know when it is necessary to move people from the community into supportive housing for safety or service-continuity purposes.

Nursing Home Care

The rate of nursing home use has doubled since the introduction of Medicaid and Medicare in 1966. Although only 5 percent of the age 65+ population is institutionalized at any one time, women have a 52 percent and men a 30 percent probability of being admitted to a nursing home in their life (Cohen, Tell, and Wallack, 1986). Current predictions of nursing home bed need do not take into account any significant shift to assisted living housing. Currently, growth is predicted to increase from 1.5 million beds in 1990 to 2.6 million in 2020 (U.S. Senate Special Committee on Aging, 1991).

Social and environmental factors continue to be instrumental in avoiding institutionalization. The most influential factors that promote avoidance of a nursing home stay are the availability of informal care (spouse or family), a lack of nursing home beds in the community, access to home care, the high income and financial status of the older person, and their personal resolve to stay in the community (Rivlin and Weiner, 1988).

Advocates for the expansion of home care, assisted living, and adult day care argue that nursing home expenditures can be reduced when formal-informal care partnerships are employed. Experiments like **On-Lok,** in San Francisco, and **Rosewood Estate**, in Roseville, Minnesota, routinely report cost of care ratios that are 20–25 percent lower than conventional nursing home costs, while offering a much higher quality of life. However, large-scale policy studies conducted to test the overall cost of alternative long-term care systems are frequently inconclusive.

Latent Demand?

The major unknown in these studies is the potential latent demand represented by the approximately 5 million older disabled elderly who are currently cared for in the community by spouses, family members, and informal caregivers (Rivlin and Weiner, 1988). Would these people give up taking care of relatives and friends if a more palatable long-term care alternative were available? In the current system, the nursing home is considered such an undesirable alternative that placement is accepted only after all other options have been exhausted. In order to reduce formal long-term care costs, we have made the worst alternative the only alternative.

In northern Europe, national health insurance has made the provision of community-based care easier for many families to manage. Its approach has involved a range of community alternatives such as home care, respite care, adult day care, and service homes. The most acutely ill and disabled are placed in settings similar to subacute hospitals in the United States. Nursing homes are being phased out in place of a system that keeps the frail in their own homes or purpose-built residential environments, and places only the most acutely ill in hospital-type settings.

6. WHAT CONTRIBUTION DOES ARCHITECTURE MAKE IN FURTHERING THE GOALS OF ASSISTED LIVING?

Introduction

Architecture plays two important roles. The first involves its impact on purpose and function. This can be evaluated by measuring how well rooms encourage social exchange and how effectively equipment, fixtures, surface materials, room layouts, plan diagrams, and furnishings facilitate accessibility and environmental manipulation.

The second role is one influenced by symbol and association. This is the most overlooked but potentially most powerful influence affecting behavior and attitude. The imagery and appearance of the environment establish how a setting is perceived by residents, visitors, relatives, and staff. If a setting is viewed as an institution, then residents are often seen as sick, feeble, or unhealthy. Viewed from the perspective of a residential environment, expectations regarding competency, independence, vulnerability, and dependence are often different. The environment becomes a frame that establishes expectations and beliefs.

This is true of other forms of architecture that have normative associations, like schools, churches, and hospitals. In these settings, "function follows form." That is, the form of the environment establishes strong expectations for behavior and use. Another reason to avoid the institutional appearance of the nursing home, beyond the issue of unattractiveness, is the counterproductive feeling it creates regarding residents and their competency.

Considerate and Challenging Architecture:

The two opposing characteristics of support and challenge should be addressed in the layout of community spaces and the specification of wall, ceiling, and floor materials. The environment should be designed to encourage social interaction but also assure privacy. Residents should be challenged to exercise arm and leg muscles, but the dining room should be within a comfortable walking distance. Open stairs should be available to encourage residents to walk, but elevators must be accessible to all. The ability to exercise choice allows the older resident to balance these two opposing positions. The environment must facilitate flexibility and choice so residents can make decisions based on individual competency and personal need.

FIGURE 2.4 Stairs are important architectural elements that lend character and efficiency to a housing arrangement. At the *Elder Homestead* project, in Minnetonka, Minnesota, a bench seat was added to the stair landing so older residents can rest. Additions like this, allow the stairs to serve as a therapeutic exercise for a range of residents.

Stimulation, Facilitation, and Manipulation

Sensory stimulation, social engagement, and opportunities for passive observation are also important to facilitate. Environments that utilize different landscape materials take advantage of natural light, use color creatively, add to spatial differentiation and stimulate the senses. Settings that encourage informal social exchange and friendship formation can enrich life and overcome depression. Places that allow residents to vicariously engage in watching on- and off-site activity will stimulate the mind and spirit.

Finally, doors and windows that can be easily opened and manipulated, shelves that can be reached, fixtures that can be safely maneuvered, work activities that can be carried out with ease, graphics that can be read and understood, and activity spaces that can be reached easily are important dimensions of the functional environment. These can hamper the older person's independence if they are not properly addressed. Environments for the aged must recognize the vulnerability of the oldest-old. Younger elderly with greater upper and lower body strength, better balance control, normal aerobic capacity, and better capacity for homeostasis can overcome barriers that may stop the frail older person. The physical environment must not create barriers that limit choice, control, independence, and autonomy.

7. WHAT ROLE CAN THE FAMILY PLAY IN HELPING THE OLDER FRAIL PERSON?

Introduction

One of the major problems with conventional nursing home facilities in the United States is their attitude toward inviting and accommodating family members in the institution. One major impediment to family interaction is the small semiprivate room most residents share. Private conversations and personalization of the environment suffers when a room is small and semiprivate. Families feel unwelcome and uncomfortable in nursing homes because there are few common spaces that accommodate family interactions. Spaces like small cafés, intimate sitting areas, solarium greenhouses, picnic spaces, and enclosed porches are absent. This lack of inviting common space gives families the feeling they are not an important constituency, reinforcing the dominant nature of management's presence.

FIGURE 2.5 *Nybodergaården,* in Copenhagen, Denmark, like many other Danish nursing homes, contains a bar on the first floor where mixed drinks, beer, and snacks are sold every afternoon. Walking into a nursing home and hearing convivial laughter and the tinkle of ice against a glass is a strange but satisfying experience. It makes one instantly recognize the latent potential for interaction that exists in long-term care facilities in the United States.

Family Versus Institution

Moving to a group living arrangement from a single family home should be considered a positive move. In an assisted living environment, safety, accessibility, and security are greater. Help is available to manage problems and new acquaintances can be easily made. However, most older people view this move as a failure. Group living seems more regimented and stigmatized. The ambiguity of fitting a past family life into this new context further confuses the situation. In nursing homes, an older resident's normal relationship with family members is often altered in radical ways, making it difficult to resume normal relations.

The majority (63 percent) of older people in nursing homes have children and therefore have the potential to develop constructive positive emotional relationships with family members (U.S. Senate Special Committee on Aging, 1991). In addition to the place being made attractive to families and being designed to support social relationships, it can also serve as a community resource. Information, education, peer group discussions, adult day care, respite care, and home care referrals are services that can benefit families.

Family Based Activities

There are many important roles family members can play within the setting and a number of activities that can enhance communication between the older resident and a family member. The following can be facilitated by the design of the building and encouraged by management policy.

1. taking meals together
2. doing personal laundry together
3. reading books
4. volunteering in group activities
5. listening to music
6. taking a picnic lunch

Sharing responsibility with the staff for instrumental activities of daily living is another approach that allows families to forge a creative partnership involving joint responsibility for care management and service provision.

8. HOW DOES REGULATION CONSTRAIN INNOVATION, EXPERIMENTATION, AND PROGRESS?

Introduction

In the United States, the regulation of group living environments has been generally overly prescriptive and rigid. The response to problems of abuse in nursing homes has been the introduction of laws and requirements that narrow flexibility and limit experimentation with promising therapies or creative approaches to care management. This heavyhanded response to the problem has created a fear-based system of accountability that undermines professionalism. It also emphasizes punitive actions over positive reinforcement or constructive feedback. The system's prescriptive nature often assigns a standard response to a problem rather than encouraging a rational and balanced assessment of the situation as the basis for potential solutions.

The most troublesome regulations are those that place limitations on the family's involvement in providing care and those that discourage therapies involving residents in daily work activities such as setting the table, washing dishes, and preparing food. State regulations require certain services in a "one size fits all" approach. This often requires staff to carry out tasks such as feeding, toileting, and bathing, and may not involve the older person in a program that challenges self-maintenance abilities. These practices undermine self-confidence, create patterns of dependence, and lessen resident competency. Keeping residents from practicing activities of daily living and taking away the option to make choices about the help they need or want is psychologically devastating.

Building Code Compliance

Regulations often center on physical design considerations because they are tangible and controllable and because problems brought about by abusive personnel are more difficult to address. Fire safety regulations often ignore new fire fighting techniques and obsess on issues of safety to the exclusion of other quality-of-life factors. Codes can eliminate the use of residential materials, limit open stairs and fireplaces, specify overly wide doors and corridors, and require solid wall separations between corridors and common rooms. These issues may seem trivial, but combined with relatively low construction budgets they often lead to stark, inhumane, and overly institutional settings. Regulations can address functional considerations such as fire exiting more easily than character and appearance concerns. Statutes emphasize "measurable" functional issues over subjective concerns such as what the place looks like.

FIGURE 2.6 European fire code provisions are substantially less restrictive than United States codes. For example, in the *Vanhainkoti-Parvakeskus Himmeli* home for the aged, in Pori, Finland, a fireplace located at one end of a widened corridor accommodates residents around the hearth.

Zoning Codes

Buildings in the United States rarely serve older people in the surrounding neighborhood through mixed-use programs that include restaurants or rehabilitative services. Providers may recognize the need to do this, but strict zoning laws specify separate land uses, and environmental impacts regarding neighborhood traffic generation and parking often force sponsors to drop these efforts after initial discussions with city officials. The result is a continuing trend toward privatization and segregation.

Zoning laws, state regulations, and building codes have combined to create a morass of overlapping restrictions that keep United States projects from emulating some of the most successful qualities of European developments.

9. WHAT ELEMENTS CONSTITUTE A THERAPEUTIC ENVIRONMENT?

Introduction

As a therapeutic model, the environment should stabilize and build the competency of individual residents. The literature contrasts environments that are prosthetic with those that are therapeutic in nature. The prosthetic environment overcomes problems by providing the older person with a device that substitutes for a disability. A grab bar adjacent to the toilet and a pair of eyeglasses are good examples. The therapeutic environment, on the other hand, provides a measure of challenge along with support that builds competency. A small kitchenette where a resident can prepare a snack exemplifies such a therapeutic feature. Lawton's competence-press theory is consistent with this idea (Lawton and Nahemow, 1973; Lawton, 1980). This theory suggests that the best match between the environment and the individual is one that engages abilities rather than passively supports need.

Physical Stimulation

Implied in the notion of the therapeutic environment is the desire to stimulate physical, mental, and emotional abilities. Physical stimulation primarily involves exercise and physical therapy. The design of the environment should provide ways to challenge physical ability. For example, an open stairs in a convenient and visible location can encourage residents to use it for exercise instead of passively relying on an elevator. Convenient access to occupational and physical therapy equipment and personnel can also encourage the older person to pursue these activities. The layout of the building can also encourage exercise through the placement of equipment in convenient central locations or along popular circulation corridors.

Family Support

Family members can offer the strongest source of emotional support and understanding. A family history of past relationships can provide a foundation of love, understanding, and affection that transcends current circumstances. The environment must support family gatherings, shared activities, and meaningful conversations between residents and family members. Finally, engaging residents in activities of daily living allows them to practice independent behaviors that meet their needs, thus creating a sense of self-reliance, and self-respect. A truly therapeutic environment should offer a range of opportunities that stimulate the mind, the spirit, and the body.

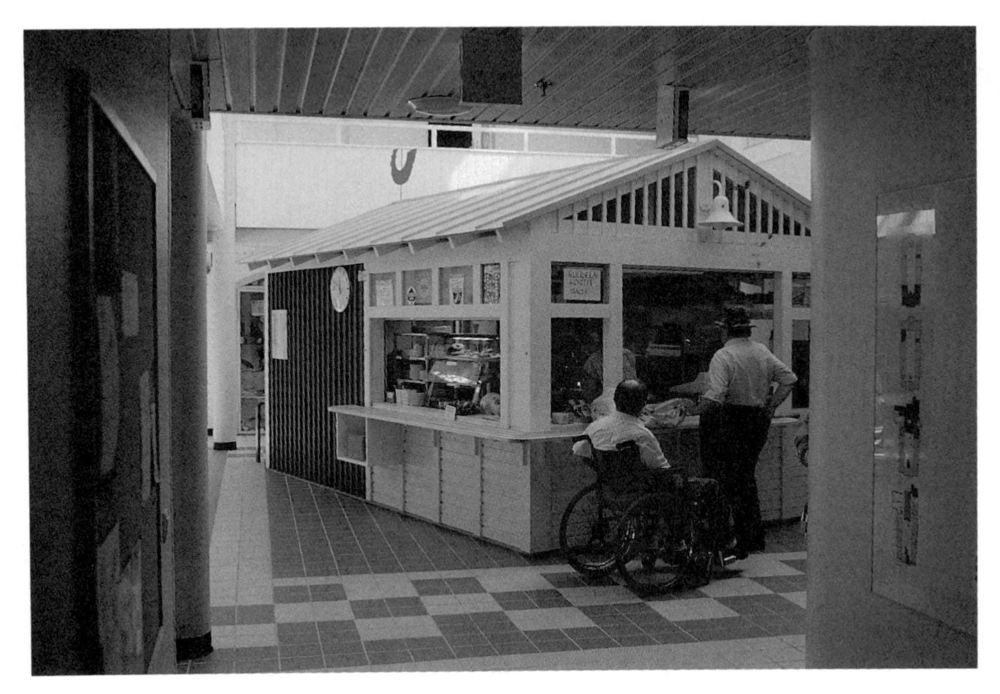

FIGURE 2.7 This kiosk located adjacent to the entry of the *Kuuselan Palvelukoti,* in Tampere, Finland, is designed to resemble the ice-cream kiosks found in urban parks in Finland. It sells snacks and has become a popular destination for residents. An adjacent lounge overlooks the garden atrium below. Residents can watch the activity generated by the kiosk and facility entrance, or passively read a book.

Mental and Emotional Stimulation

Mental stimulation and emotional support can be encouraged by ensuring access to intellectually stimulating activities, such as card playing, reading, music, and puzzles. The placement of television sets in public lounges encourages passive behaviors while discouraging conversations among residents. Television sets are best located in separate rooms or left for private viewing in a resident's room. Environments that stimulate the senses are also important. Colors, spatial configurations, surface textures, and smells offer a range of possibilities. Artwork that presents familiar positive emotional relationships can stimulate memory by recalling positive past associations. Friendships and informal socializing can ameliorate depression, giving residents a way to articulate problems and share concerns with others. Sitting areas where residents can engage in observation of onsite and off-site activities and outdoor garden areas that provoke a sense of spirituality use aspects of the environment to stimulate vicarious engagement. Finally, animal-assisted therapy facilitates emotional responses through the unconditional affection that animals provide.

10. IS INSTITUTIONALIZATION INEVITABLE?

Introduction

One of the major ways to improve our system of housing for the frail includes eliminating the habits, beliefs, and past experiences that restrict our thinking and narrow our vision. The United States system of housing and supportive services is tied to regulations that force us to think in terms of buildings not services (Tilson, 1990). This one feature has done more than anything to keep society from pursuing ideas about aging in one place. Europeans, in contrast, think of group housing as a service and not a building type. Rarely do building regulations in Denmark or Sweden force older people to move to nursing homes as they decline in ability.

Incontinence, Disability, and Dementia

In the United States, when competency declines, we assume that an older person must be moved to a health care institution. In the past, we have used minimal criteria to rationalize placement in a nursing home. Incontinence was a popular criterion for institutionalization until new technologies, products, and training protocols became popular and common. When this happened, it exposed the questionable logic of this tacit requirement. In the past, physical disability has also been used as a criterion for institutionalization, under the assumption that anyone needing a wheelchair to ambulate could not safely stay in his or her own home. Expanded home care and the legal ramifications of the American Disabilities Act have called into question the wisdom of this thinking. Currently, the same bias exists toward dementia victims. Many argue a lockup ward in a nursing home is the only place that can accommodate demented wanderers. Again, experimental prototypes with new technologies have exposed the narrow and prejudicial nature of this point of view.

Nursing Homes as Subacute Hospitals

European housing policies for the frail have in the past ten years been oriented away from institutional solutions for the sick and impaired. Most policies are centered around keeping people at home or, when frequent help and support become necessary, moving them to a service house or group home. European nursing homes have been slowly transformed into subacute hospitals that care for only the most medically indigent population. The majority of the frail live in independent arrangements that continue to challenge their competency by engaging them in a range of self-maintaining activities.

FIGURE 2.8 New technologies like the United States–based Lifeline system and the Swedish-based *Tele-Alarm,* are allowing older people to safely stay at home. Many European service houses take responsibility for monitoring community alarm systems linked to resident units, adjacent "lean-to" housing, and independent dwelling units in the surrounding neighborhood.

Home Care and Technology

Older people today are far more reluctant to accept placement in a nursing home without exhausting all possibilities. The expansion of home health care has demonstrated other choices and possibilities. However, in the United States, options for group living are not treated as entitlements like placement in a nursing home. The expansion of home care in the community and the development of sophisticated options that employ advanced communications technology and robotics will give the upcoming generation of older frail more options than in the past. These new technologies will allow older people and their families to avoid institutional placement longer.

Existing nursing homes should be more like hospitals and the need to enter them treated as the last alternative when all other options have been exhausted. The necessity of institutionalization in a nursing home should be questioned just like the "required hospital stay" has been during the last five years, under hospital utilization guidelines.

11 . HOW CAN FAMILIES AND INSTITUTIONS WORK TOGETHER TO SUPPORT THE NEEDS OF THE OLDER FRAIL PERSON ?

Introduction

A combination of informal help supplemented by home delivered services currently allows the majority (62.5 percent) of the 85+ disabled elderly population to live independently in the community (Rivlin and Weiner, 1988). Daughters, who traditionally have been a major source of informal care, are increasingly burdened. Studies, however, have been unable to link increasing labor force participation with reduced levels of caregiving to aging parents. Daughters who work balance conflicting demands around their work schedule and often sacrifice free time. Continuing demands will confront the "sandwich generation," which has caregiving responsibilities to children *and* parents. In the future, the heavy burden of caring for two generations of parents will present serious challenges to the family caregiver.

An increase in three-generation living arrangements is not likely to be a popular solution. Only 18 percent of the older population who have children choose to live with them. This percentage has stayed relatively constant during the last ten years. Furthermore, current statistics show only 12 percent of the elderly who are living with children or relatives cite poor health as the main reason (Crimmins and Ingegnari, 1990). Their presence in the three-generation family is often an aid, not a burden.

The round-the-clock needs of the most severely disabled elderly are not well met by home care personnel who make client visits in minimum four-hour time blocks. Home care help is rarely available when the older person needs toileting assistance at three in the morning.

Care Partnerships Needed

More creative alternatives for sharing caregiving responsibilities between families and formal providers must be tested. Respite care, adult day care, community care, and assisted living have the capacity to combine formal service provision with informal help. Case management techniques that establish working relationships with families like those of **Rackleff House** and **Rosewood Estate**, have the potential to minimize barriers between formal providers, families, and older persons. These approaches often challenge older people to do as much as they can for themselves, while accommodating the participation of family members in the care plan.

The term *managed risk* is used to describe a concept that allows older people and their guardians to make decisions about the nature and level of supervision they need. In most states, regulations establish this standard of care and through licensure vest institutions with the legal responsibility for implementation. Older residents and their families have little authority or power to influence these rules. One of the major differences between home care and licensed group living arrangements is the lack of personal freedom and choice created by the often rigid prescriptive rules and regulations applied to group living settings.

12. HOW CAN THE ENVIRONMENT ENHANCE INDEPENDENCE AND PRIVACY?

Introduction

Privacy and independence represent two important housing considerations that are poorly addressed in skilled nursing homes. One of the most disturbing aspects of the nursing home for visitors is the public nature of care provision and treatment. Care is rarely provided behind closed doors. Furthermore, in nursing homes the boundaries between private and public domains are confused. The vast majority of arrangements require residents to share a small room with another unrelated person.

The line between dependence and independence is also fuzzy and may differ for each person. It can change from day to day depending on how people feel. Rigid regulations can drive caregiving by requiring services that don't consider an individual's ability to help himself. Oversupport erodes competency by not allowing residents to participate in self-care and self-maintaining behaviors.

Regulations Affect Solutions

Solutions to privacy and independence problems exist but are ignored for a variety of economic and regulatory reasons. For example, providing residents with private single-bed rooms can't be achieved efficiently without violating maximum distance criteria from a nurses' station to a patient's room. The total life cycle costs associated with nursing home care over the forty-year life of a building are very large in comparison to the difference in cost between a single and double room. The vast majority of costs (60–75 percent) are tied to labor costs. The total building construction costs account for about 8 percent of life cycle costs. In other words, costs associated with making rooms bigger and more private are a tiny fraction of total life cycle costs, but the added staffing, which is linked through regulation to the number of accessible nurses' stations, forms a major stumbling block (Hiatt, 1991). Although maximum distances from a nurses' station to a resident's room were established to minimize staff time spent in transition, little work has focused on studying alternatives such as decentralized staffing schemes and the true costs of these distance criteria.

The design of a typical nursing home room also does not support independence. Toilets located in corridors or shared with other patients further minimize self-care options. In most nursing homes the idea of minimal food preparation facilities within a resident's unit is considered entirely inappropriate. Thus, self-maintaining strategies that involve the family and challenge the older resident to do as much as possible for herself with limited assistance from staff are impossible to implement. Independence is greatly affected by physical strength, cognitive ability, and emotional stability. Therapies that strengthen these competencies or reduce their decline are important components of any program centered on enhancing independence.

13. SHOULD THE MENTALLY AND PHYSICALLY FRAIL BE INTEGRATED OR SEGREGATED?

Introduction

In the early stages of organic mental disorders, memory lapses, confusion, and temporary disorientation are relatively common. Assisted living environments should expect to attract as many as 40 percent of their residents with these types of problems. Surveys show more than 60 percent of the nursing home population suffers from disorientation or memory impairment (U.S. Senate Special Committee on Aging, 1991). When an individual becomes difficult to manage because of constant wandering or when a resident's lack of emotional control leads to screaming and inappropriate behaviors, coexistence with less mentally impaired residents becomes a problem.

The social and communal nature of assisted living can make integration difficult. A major strength of assisted living involves the relationships that residents have with one another and their ability to create an active social life. This can be threatened when a resident is confused or can no longer participate in a socially competent way. Thus, social integration, which is the strength of the assisted living housing model, becomes a liability when the heavily mentally frail comingle.

In European housing and in some United States-based assisted living arrangements, dementia victims are moved to separate floors or wings of the facility. In Sweden, small group homes for dementia victims have been created as a component of the housing continuum. Programs like adult day care and special therapies can also be pursued within a group home. In these settings, older people live in an intimate housing arrangement that allows wandering, activity of daily living therapies, and privacy, to be provided within a controlled residential setting.

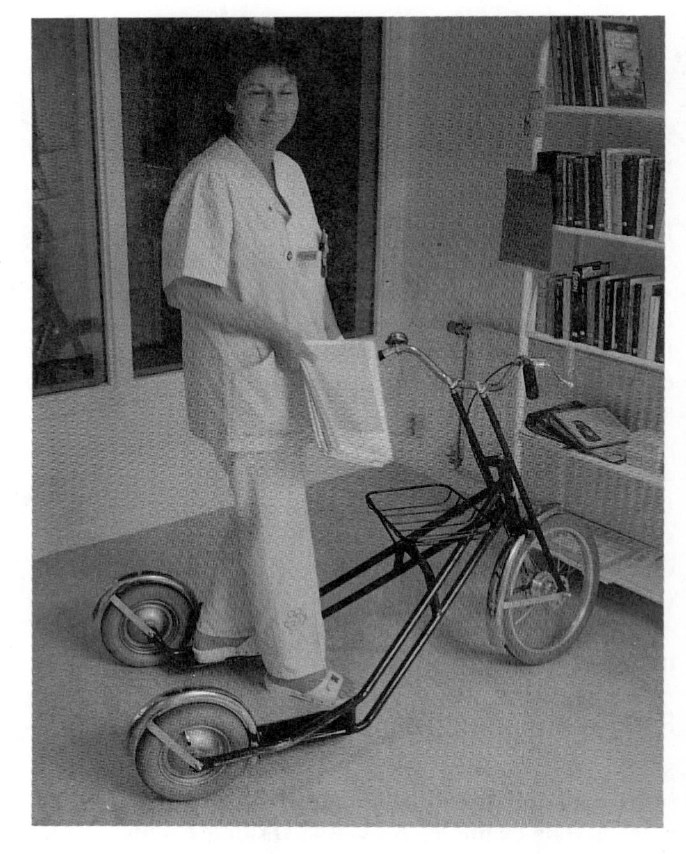

FIGURE 2.9 Danish nursing homes do not follow the same regulatory code restrictions as American facilities. The result is a greater range of unit configurations. During nightshifts when nurses make rounds between decentralized wards, they utilize scooters to get from one place to another in less time.

14. HOW CAN COMPASSIONATE AND COMPETENT STAFF BE RECRUITED, TRAINED, AND RETAINED?

Introduction

One of the most important aspects associated with successful management is vesting competent and motivated employees with the authority and responsibility to make decisions about resident care. Attending to the needs of frail residents should be based on understanding how to motivate positive behaviors. Personnel must be keenly aware of the needs of the older person, and continually think about how to motivate them to do things for themselves. In a sense, their job is like a schoolteacher's or coach's; they must know how to motivate people to bring out their personal best.

Decentralized management allows this form of personal contact to work best. Having an attractive noninstitutional environment within which to work can affect motivation, giving staff the feeling they are implementing a philosophy of care that places the resident ahead of other priorities. When the appearance of the environment is not institutional, it can stimulate innovation and creative problem solving.

Staff Incentives

Designing spaces for staff to relax during breaks and between shifts is also an important symbolic message management can provide to workers. However, a host of other factors has stronger impacts on staff performance than the environment. Selecting motivated people with true affection for older people is an important hiring prerequisite. Allowing employees a flexible work schedule that can be managed around child care or other family obligations is important to some. Providing training that leads to a deeper understanding of the behavioral side effects of aging allows the staff to better understand the human conditions underlying aging. This makes workers more cognizant of their own aging and the aging of family members. Finally, discount meals, stock options, and compensation directed toward building a successful organization are also important. The staff must recognize the importance of their contribution to the overall success of the organization.

Compassionate, caring, and thoughtful staff behaviors may ultimately rest on the leadership models provided by administration and the organization that sets policy. The environment is like an instrument that can narrow or boost the achievement of the staff. Saddled with a substandard physical environment, a highly motivated staff will find their efforts thwarted. A carefully designed setting allows creativity to soar and the spirit of the possible to overtake pessimism.

FIGURE 2.10 Group homes for dementia patients are often developed in conjunction with service houses to allow older demented persons to stay in the same general setting as they age and become more frail. The kitchen of this group home in the *Kuuselan Palvelukoti,* in Tampere, Finland, is located on the top floor of the service house and accommodates eight residents.

SUMMARY

The need to change current policies that rely on the nursing home to care for older frail people is a concern based on humanity and economics. It is cheaper for society and better for older people to have professionally managed group living arrangements that are residential in nature. Trends in the United States clearly point to the replacement of the nursing home with other alternatives that ensure greater privacy and encourage independence.

The fourteen questions in this chapter deal systematically with issues of cost, demand, architectural design, family support, regulatory interference, therapeutic intervention, and staff recruitment. Assisted living is forcing a radical rethinking of our values, assumptions, rules, and practices. Middle-aged Americans today view the nursing home as an unacceptable solution to society's long-term care problems. We must begin now to question the purpose and meaning of institutional practice and devise new ways of helping older frail people lead more satisfying, vital lives.

DEFINITIONS AND PRINCIPLES 3

INTRODUCTION

Creating an innovative and attractive assisted living environment requires careful thinking about how the building will serve the needs of residents, promote therapeutic goals, efficiently operate, encourage social exchange, and support stimulating activities. The design process is a complex, often disjointed and sometimes unpredictable one that requires constant attention to the broader issues and objectives established for each project.

This chapter reviews methods and techniques for controlling and guiding the design development process, identifies important behavioral objectives and project qualities, and proposes defining characteristics of assisted living that aid concept development and project refinement.

ORGANIZING THE DESIGN PROCESS

In order to guide the design and project development process, one must first establish a methodology that:

1. **Identifies** salient behavioral objectives and project design principles prior to commencing design;
2. **Reviews** important design ideas during design and construction, in light of specific project considerations (budget, community context, codes, program needs, operational objectives, and resident preferences), and
3. **Evaluates** the project after it is completed and occupied.

Identifying basic assumptions during the design and development process keeps important considerations from being forgotten or overlooked. It also permits the building development process to take on characteristics of an experimental research design. Structured in this way, each building becomes an experiment complete with design-behavior hypotheses and belief statements that can be tested after occupancy. When a process like this is not employed, projects may emerge after design and construction without a clear sense of purpose.

For a more complete discussion of design research methods and a more comprehensive treatment of the programming, design review, and postoccupancy evaluation process, please refer to John Zeisel's book *Inquiry by Design* (1981).

FIGURE 3.1 One of the environment-behavior hypotheses at the *Captain Eldridge Congregate House,* in Hyannis, Massachusetts, was that management be accessible and visible, but not overly controlling. A subtle relationship was developed by placing the manager's office on one side of the entry foyer. It is linked to the foyer by a Dutch door and a double-hung window.

ESTABLISHING OBJECTIVES

Perhaps the most effective way to ensure that behavioral considerations will be inspected during the design process is to develop an architectural space program that anticipates how spaces will affect resident behaviors and activities.

Behaviorally Based Architectural Program

This type of architectural program identifies the activities, behavioral objectives, operational influences, and design considerations for each space, in addition to the square footage, space description, and adjacency notations found in more conventional architectural programs (Preiser, 1978). In a behaviorally based architectural program, these descriptions for each space establish expectations and performance goals for the building. During the schematic design phase, they are reformulated as criteria and used to test ideas about architectural form and organization. After occupancy, they become a "user's manual" for the building, allowing management to note the intentions, activities, and behavioral outcomes each space was designed to accommodate. Another important advantage of this type of program is its ability to inform the client about intention and to reach a shared understanding of how spaces will support various activities and patterns of use.

Design-Behavior Hypotheses

Another design review method that establishes a basis for a shared understanding of spatial purpose and intention is the design-behavior hypotheses. In his work with the **Captain Eldridge Congregate House**, Zeisel identified nineteen design-behavior hypotheses that formed the basis for many of the ideas explored in this project. They deal with a range of important influences. For example, an open stairs placed near the center of the project was designed to entice residents to take the stairs, instead of an adjacent elevator, for physical health and exercise purposes. A management office located off the small entry lobby was designed to ensure security, but exert a low-key management influence on the project. The hypotheses displayed in Chapter Five are fashioned to clarify design intentions while allowing them to be evaluated during the occupancy stage.

Design Review Questions and Design Directives

Two other techniques that can be used to further test the quality and validity of alternative solutions as they are developed and refined are design review questions and design directives. Design review questions (Zeisel, Epp, and Demos, 1977; Zeisel, Welch, Epp, and Demos, 1983) raise specific questions about the ability of a design scheme to support activities, accommodate resident needs, or further the behavioral objectives established for various spaces. Normally, these are listed for specific spaces and can be referenced during the process of design refinement to ensure the design accommodates various functions. Design directives (Regnier, 1985; Regnier and Pynoos, 1987) are findings from design research activities that reveal conflicts and problems, as well as suggest good practice solutions, which can enhance project performance.

Design Guidelines

Finally, statements of good practice are frequently employed in manuals that seek to provide comprehensive advice about design decisions (AIA Foundation, 1985; Valins, 1988; Carstens, 1985; Green, Fedewa, Johnston, Jackson, and Deardorff, 1975; Lawton, 1975; Raschko, 1982). These often combine timeless good practice experience with research findings. The purpose is to create parameters for a specific design problem. For example, designing outdoor spaces requires an understanding of plant ecology, sun and shade patterns, social behavior, and design aesthetics.

Behaviorally based programs, design-behavior hypotheses, design review questions, design directives, and design guidelines are all useful and effective methods to enhance design quality. They help by clarifying the problem better and identifying in advance unintended negative side effects of various solutions.

Physical, Social, and Organizational Context

Employing design methods that continually remind the design decision maker of the side effects associated with various decisions is more important in a housing type like assisted living, where operational objectives, therapeutic considerations, and the desire to enhance social exchange are present. In their work with Alzheimer's residents, Cohen and Weisman (1991) developed a model for understanding relationships between the physical, organizational (facility policies, caregiving, and management), and social (residents, visitors, friends, and relatives) contexts of a facility (Chart 3.1).

In their conceptualization, the physical environment is recognized as having impacts on both the social and organizational domains. Cohen and Weisman suggest that these three domains must be coordinated with one another in an effort to facilitate therapeutic goals, which come from an understanding of the special needs of mentally frail people. Although this model was developed to describe relationships in dementia facilities, it is relevant to assisted living environments. Its implementation requires a clear understanding of what constitutes "therapeutic goals." To understand this further, it is useful to explore what other researchers have identified as the most salient guiding principles for designing assisted living.

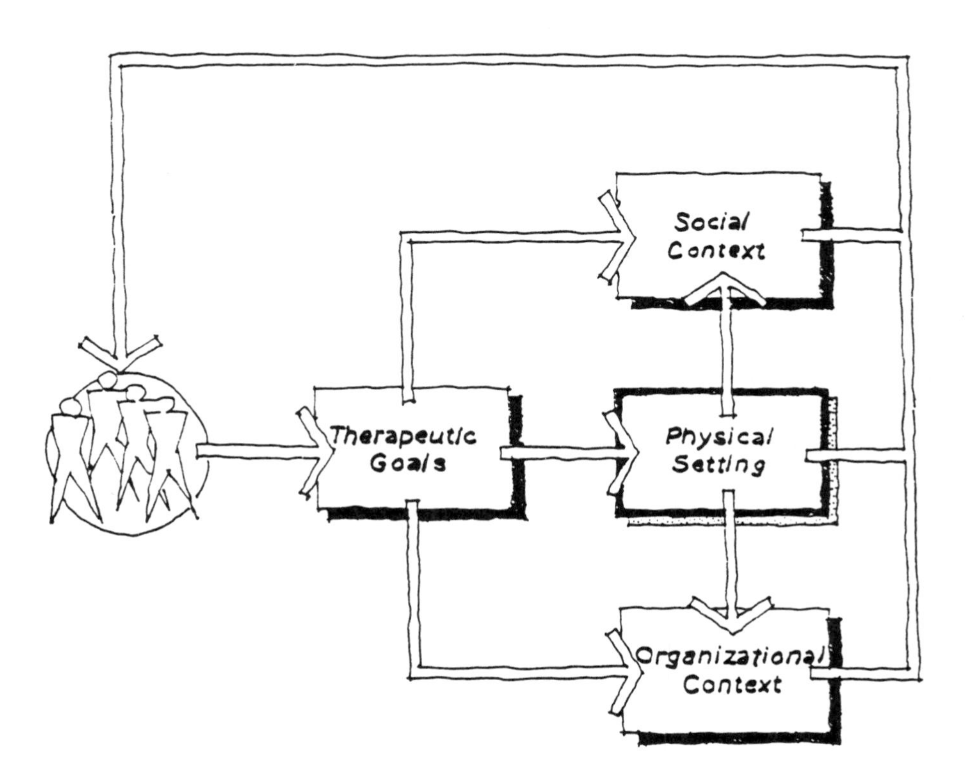

CHART 3.1 This conceptual framework suggests that facility goals must be implemented by examining their relationship to the social, organizational, and physical context of the environment. The physical environment also greatly influences the social context (relationships between residents, family, and friends) and the organizational context (facility policies and management interaction) (Cohen and Weisman, 1991).

DESIGN PRINCIPLES

Regnier and Pynoos Common Research Themes

In assembling the seventeen research projects used as the basis for the 1987 publication *Housing the Aged: Design Directives and Policy Considerations* (Regnier and Pynoos, 1987), it was clear a number of the research contributions focused their inquiry around six distinct themes. These themes created a partial list of priority areas that researchers had addressed through their work and formed the basis for a discussion of future research directions.

1. **Resident Satisfaction:** Studies oriented toward measuring life satisfaction, quality of life, or other attributes of resident preference and desire as they relate to physical design and operational policies of housing for the aged.
2. **Social Interaction:** Studies oriented toward understanding more about how physical and operational environments stimulate, enhance, protect, and nurture informal and structured social exchange.
3. **Management:** Studies that identify management as an important influence in creating or quashing activities and behaviors that enhance resident quality of life.
4. **Sensory Aspects:** Studies that address how the environment can respond to the changing sensory modalities of the older person in ways that compensate for aging losses.
5. **Physiological Constraints:** Studies that examine grip strength, reach capacity, muscle strength, motility, and ambulation ability in an effort to better understand how the environment can support these losses.
6. **Wayfinding:** Studies that examine how older residents orient themselves within buildings and also suggest how to design settings to allow residents and visitors to effectively find their way around them.

Wilson's Assisted Living Concepts and Attributes

In clarifying her model of assisted living, Keren Brown Wilson (1990) recognizes six *attributes* of the physical and operational environment and identifies four *concepts* that underlie her management philosophy for housing the frail elderly. The attributes include:

1. **Privacy**
2. **Dignity**
3. **Choice**
4. **Independence**
5. **Individuality**
6. **Homelike Surroundings**

The following four concepts are accompanied by suggestions about how to implement them through a consistent management philosophy. They include:

1. **Create a Place of One's Own:** Assure privacy through a locked door, a private bathroom, and the ability to prepare food.
2. **Serve the Unique Individual:** Each resident's needs and abilities are different. Recognizing these differences is the basis for an effective personal therapeutic strategy for each resident.
3. **Share Responsibility Among Caretaker, Family Members, and Resident:** Creating care partnerships between formal providers and the family gives everyone, including the older person, the option of participating in care management.
4. **Allow Resident Choice and Control:** Simplifying choices can expand the number of decisions residents can make. The more residents can exercise choice, the more control they have over a given situation.

Cohen and Weisman's Therapeutic Goals for Dementia Facilities

In establishing a basis for operationalizing their model of design, management, and resident interaction, Cohen and Weisman (1991) have identified nine therapeutic goals that form the basis for operational policies, physical design solutions, and social interaction. These nine goals are established for the mentally frail, but are also equally relevant for the physically frail. They include:

1. **Ensure safety and security**
2. **Support functional ability through meaningful activity**
3. **Heighten awareness and orientation**
4. **Provide appropriate environmental stimulation and challenge**
5. **Develop a positive social milieu**
6. **Maximize autonomy and control**
7. **Adapt to changing needs**
8. **Establish links to the healthy and familiar**
9. **Protect the need for privacy**

FIGURE 3.2 Danish nursing homes keep a medical log and a social log that chronicle patient activities and status. The social log is a diary, which describes patient status and is used to communicate information to the next nursing shift. Both nurses and patients feel free to write in this journal. This is a great example of how the Danes have transformed a recordkeeping device into a social communication tool.

Regnier and Pynoos's Environment-Behavior Principles

The following composite list of twelve environment-behavior principles combine and expand the categories in the above lists (Regnier and Pynoos, 1992). Although all of these are relevant for older people living in housing and service arrangements like assisted living, many can also be applied to older frail people living in home environments or nursing homes. Included on the list are concepts like privacy and social interaction that are conceptually in opposition to one another. All environments, however, should support a range of options for social engagement and privacy. Thus, we expect the environment to satisfy both ends of the social interaction-privacy continuum. Another pair of opposed principles is safety/security and stimulation/challenge. In developing a balanced individual housing context, we must recognize that each person has a need for support and a need for challenge.

Other aspects represented on this list are timeless qualities of stimulating architecture relevant to all populations. However, in many cases, the special needs of the elderly for a more accommodating architecture further underscore their meaning and importance. Orientation and wayfinding is a good example. Most people have difficulties in complex environments where few cues are available for navigation. However, older frail people with spatial memory impairments are more easily confused by a symmetrical plan and less able to differentiate similar residential floors from one another.

Another important influence is operational policies, which can reinforce or reduce the effectiveness of each principle. For example, a lock on the door does little for privacy when staff members have pass keys and do not knock before entering. Operational policies and environmental design attributes must be considered as complementing and affecting one another.

Paraphrased and adapted from an earlier article (Pynoos and Regnier, 1991), the following twelve design-behavior principles suggest ways in which they can be applied to assisted living environments by describing their salience to the older frail population.

FIGURE 3.3 This enclosed street in the small town of Nr. Nebel, Denmark, is located adjacent to a well-traveled street. Residents gather at a table that overlooks the nearby street activity and serves as an excellent setting for observation and social interaction.

1. **PRIVACY**
   Provide opportunities for a place of seclusion from company or observation where one can be free from unauthorized intrusion.
   This is important because it provides the older person with a sense of self and a separateness from others. Auditory and visual privacy are important subcomponents of physical separation. Privacy is more difficult to ensure in group living arrangements. Nursing home settings that rely on double-occupancy rooms severely limit access to privacy.

2. **SOCIAL INTERACTION**
   Provide opportunities for social interaction and exchange.
   This is important because one of the most important reasons for creating age-segregated group living arrangements is to stimulate informal social exchange, recreational activities, discussion groups, and friendship development. Social interaction counters depression by allowing older people to share problems, life experiences, and daily events.

3. **CONTROL/CHOICE/AUTONOMY**
   Promote opportunities for residents to make choices and to control events that influence outcomes.
   This is important because older people are often more alienated, less satisfied, and more task dependent in settings that are highly restricted and regimented. Having a sense of mastery and control has been found to have pronounced positive effects on life satisfaction. Independence is often defined by our ability to make choices, control events, and be autonomous.

4. **ORIENTATION/WAYFINDING**
   Foster a sense of orientation within the environment that reduces confusion and facilitates wayfinding.
   This is important because feeling lost or being disoriented within a building is a frightening and disconcerting feeling that can lessen confidence and self-esteem. Older people who have experienced some memory loss are more easily disoriented within a featureless symmetrical complex environment. Signs can overcome some problems but never provides the person with the confidence of knowing exactly where they are within the larger environmental context.

FIGURE 3.4 Designing sensory gardens for the physically challenged is difficult. In this Danish project, a garden pool accommodates wheelchair-users in an active, participatory way.

5. **SAFETY/SECURITY**

   Provide an environment that ensures each user will sustain no harm, injury, or undue risk.

   Older people may experience physiological and sensory problems, such as visual impairments, balance control difficulties, lower body strength losses, and arthritis, which make them more susceptible to falls and burns. Changes in bone calcium levels with aging can also increase their susceptibility to broken bones and hips. The elderly experience a high rate of injury from home accidents.

6. **ACCESSIBILITY AND FUNCTIONING**

   Consider manipulation and accessibility as basic requirements for any functional environment.

   This is important because older people often experience difficulties manipulating the environment. Windows, doors, HVAC controls, and bathroom fixtures can be hard to twist, turn, and lift. Furthermore, older people confined to a wheelchair or dependent on a walker must have environments that are adaptable enough to accommodate these devices. Reach capacity and strength limitations are therefore important considerations in the layout of bathrooms and kitchens, and in the specifications of finishes.

7. **STIMULATION/CHALLENGE**

   Provide a stimulating environment which is safe but challenging.

   This is important because a stimulating environment keeps the older person alert and engaged. Stimulation can result from color, spatial variety, visual pattern, and contrast. Stimulation can also involve animating the setting with intergenerational activities, pet therapy, or a music program. Environments overly concerned with maintenance and cleanability are often uniform in color and pattern, noisy and disconcerting to the ear, and glaring and reflective in appearance. Each individual resident is different and should be allowed to experience an optimum level of complexity and challenge.

8. **SENSORY ASPECTS**

   Changes in visual, auditory, and olfactory senses should be accounted for in the environment.

   This is important because older people tend to suffer age-related sensory losses. The sense of smell, touch, sight, hearing, and taste decrease in intensity as a person ages. Sensory stimulation can involve aromas from the kitchen or garden, colors and patterns from furnishings, laughter from conversations, and the texture of certain fabrics. A range of sensory inputs can be used to make a setting more stimulating and interesting.

9. **FAMILIARITY**
Environments that use historical reference and solutions influenced by local tradition provide a sense of the familiar and enhance continuity.
This is important because moving into a new housing environment is a very disorienting experience for some older people. Creating continuity and connection with the past is reassuring and facilitates the transition. Residents take cues from the environment. When it is designed to accommodate traditional events and fits into the regional housing vernacular, it appears more predictable and understandable. Institutional environments often use imagery that does not come from residential references and therefore appears foreign and alienating.

10. **AESTHETICS/APPEARANCE**
Design environments which appear attractive, provoking, and noninstitutional.
This is important because the overall appearance of the environment sends a strong symbolic message to visitors, friends, and relatives about the older person. Housing that appears institutional provides cues to others about the competency, well-being, and independence of residents. Staff and caregiving personnel are also highly affected by the appearance of the physical and policy environment. Personnel working in a building that resembles a nursing home will lessen cognitive dissonance and act in ways that are consistent with what an institutional context suggests.

11. **PERSONALIZATION**
Provide opportunities to make the environment personal and to mark it as the property of a unique single individual.
This is important because it allows older residents to express self-identity and individuality. In nursing homes, individual expression is often very limited. Patients do not have much personal space in compact two-bed rooms furnished with hospital beds and over-the-bed trays. Personal items used for display and decoration are often very important and salient to the older person. Collectible items may trigger memories of travels to other countries or emotional bonds with family and friends. These items can animate a room by recalling past associations.

12. **ADAPTABILITY**
An adaptable or flexible environment can be made to fit changing personal characteristics.
This is important because older people age differentially. Some have mental impairments while others suffer visual losses. For some, chronic arthritis keeps them from performing activities of daily living, while for others, arthritis is an occasional problem that is more of an annoyance than a disabling disease. The environment has the capacity to compensate for many deficits and to adapt to changing resident needs. Bathrooms and kitchens are the major rooms in which work activities take place and where safety is a major consideration. Environments should be designed to be adaptable to a range of users, including those who need wheelchairs and walkers.

DEFINING ASSISTED LIVING

Although there are commonly assumed definitions of this evolving housing type, it is important to state qualities that all projects should strive to embrace. Although few facilities today meet all of the following characteristics, they provide a normative standard for the physical and operational environment of assisted living housing. The nine qualities are as follows:

1. Appear Residential in Character
2. Be Perceived as Small in Size
3. Provide Residential Privacy and Completeness
4. Recognize the Uniqueness of Each Resident
5. Foster Independence, Interdependence, and Individuality
6. Focus on Health Maintenance, Physical Movement, and Mental Stimulation
7. Support Family Involvement
8. Maintain Connections with the Surrounding Community
9. Serve the Frail

Dealing with each of these nine qualities in more depth provides a better understanding of how each attribute can be explored. The following expands each quality by describing more about how it can be implemented.

FIGURE 3.5 *Höfje van Staats* is a social housing project that has been continuously operating since 1730. Many generations of older people have lived here. Recently a chairlift was installed for a stroke victim who could no longer use the stairs to reach her bedroom. Housing adaptations have been creatively pursued because of an aggressive national policy to keep older people in their own homes for as long as possible.

1. Appear Residential in Character

The character, appearance, precedent, imagery, and memory, of assisted living should be related to residential housing. These associations can be explored through the appearance and configuration of the building or the furnishing of interior spaces. The outward appearance of the building should employ residential elements such as sloped roofs, attached porches, and dormers for scale and association purposes. Residential materials, finishes, and treatments should be used to clad and enclose. The building should fit into the surrounding context of neighborhood residential properties. Inside rooms should be consistent with residential proportions. Larger spaces like the dining area can be broken into several smaller separate rooms. Residential materials should be specified for interior finishes and furnishings. Rooms should have varied character and purpose, as in a typical house. Open stairs should connect floors and units should be clustered in small groups to stimulate the development of informal friendships and helping behaviors.

2. Be Perceived as Small in Size

The minimum number of units to achieve economies of scale will vary by region, state, and urban context. In a small town context one may be able to operate a project with as few as twenty-five units. However, most settings will require more than forty units to offer competitive rental rates and provide reliable twenty-four-hour care. However, the larger the building, the more it can easily overwhelm residents. When residents know one another and the administrator, a feeling of family develops. This builds a second reference group, in addition to related family members. Creating a building that resembles a big house or courtyard villa can achieve compactness in plan and solidarity in community relationships.

3. Provide Residential Privacy and Completeness

A small kitchenette and a full bathroom make the dwelling unit complete. Providing extra space for an overnight visitor or family member can be achieved in units as small as studio alcoves. Privacy should be achieved through a combination of efforts, including leasing policies that encourage single occupancy, design features such as locks on doors, and management practices that require staff to identify themselves before entering. Personalization should allow furniture, display items, photos, artwork, and special collectibles to be displayed.

4. Recognize the Uniqueness of Each Resident

Each older person who enters assisted living has lived life in a unique way. Each has a multiplicity of different experiences, which have nurtured diverse interests, abilities, and values, through the acquisition of a highly personal base of knowledge. Gerontologists like James E. Birren argue that as we age, life's experiences and our own personal strengths and weaknesses make us more unique rather than uniform in our beliefs and understanding and appreciation of daily life events. Capturing that diversity within a group setting is important. Zeisel and his colleagues (1984) describe the various resident personality profiles and their role in the life of a housing environment. They suggest how the diverse abilities and interests of residents can be the basis for programming efforts. Activity programming, participant management, and the proper nurturing of individual interests can make a setting extraordinarily rich, moving it from a setting dependent on preconceived planned group activities to one that recognizes differences through small group activities that are meaningful to participants.

5. Foster Independence, Interdependence, and Individuality

Resident assessments should inventory the unique capabilities and competencies of each person and devise a treatment plan that treats each person as an individual, with respect and dignity. This approach can clarify how the community can help the person and how the person can contribute to the community. One-sided caregiving without resident reciprocity builds dependence, not a sense of belonging to a community. Assessments should identify, preserve, and build on residents' strengths while overcoming weaknesses through therapy and prosthetic intervention.

6. Focus on Health Maintenance, Physical Movement, and Mental Stimulation

Avoiding institutionalization as long as possible is a major motivation provided by assisted living. Monitoring health through preventive checks, good nutritional habits, and careful attention to pharmaceuticals constructs a safety net of assurance. Physical challenges in the form of exercise therapy can build upper and lower body strength, increase aerobic capacity, and achieve muscle control over problems like incontinence. Spiduso and Gilliam-McRae, in a recent review article (1991), underscore the effects of disuse on human frailty by the extensive use of medications and physical restraints in nursing homes. They describe in detail the effects of exercise on the four chronic diseases (cardiovascular disease, diabetes mellitus, chronic lung disease, and arthritis) that cause the greatest disability in the elderly. Activities that stimulate the mind, like reading and discussion groups, also create opportunities for friendship formation, informal social exchange, and the sharing of personal feelings. This counteracts depression while replacing friendships that have been lost through attrition or relocation.

7. Support Family Involvement

Most institutions treat patients like a baton in a relay race. Family responsibility ends when the resident moves to the institution. Maintaining strong family connections to the resident is often difficult after institutionalization. The purpose of family-based assessments is to develop a caregiving partnership that allows family members a more important role in making critical decisions and in managing care. The building should also provide places for residents and family members to gather and share activities. Encouraging overnight stays outside the facility at the family's house or within a resident's unit can also add to family connectivity. When family members become a participating partner in the life of the place, they add vitality and energy, while saving money. They can also important to the lives of other residents.

8. Maintain Connections with the Surrounding Community

Encouraging residents to visit their old neighborhood to attend church or have their hair styled maintains linkages and connections with old friends and familiar places. This allows residents to draw on a wider range of interactions rather than narrowing their choices. Housing projects that develop inventive ways to serve the surrounding community become less internally focused and better connected to the fabric of the community. Intergenerational exchange programs with preschools have been successful in forging exchange relationships with older people and children. Foster grandparents can improve self-esteem and receive affection and admiration.

9. Serve the Frail

The average resident of an assisted living setting is likely to be in the 82–87 year age range. Furthermore, facilities should conform to the 40/40 rule, which suggests that 40 percent of the population could be experiencing some problem with incontinence and 40 percent some confusion or memory impairment. It is not at all uncommon to have as many as 80 percent of residents needing some bathing assistance and 40 percent needing toileting assistance. A population with these needs can be managed within assisted living as an alternative to the nursing home.

These nine definitional characteristics stress the need for residential appearance, privacy, independence, therapeutic intervention, family support, and connection to the surrounding community. Although not totally inclusive, this list provides a good starting place for thinking about how assisted living environments can provide the frail with a higher quality of life.

FIGURE 3.6 The Danes involve residents in a range of therapeutic activities centered around the concept of occupational, or ergotherapy. Weaving is one of the most popular activities because of the positive therapeutic value it provides finger, leg, and arm muscles. The rugs and tablecloths produced through these efforts are handsome and instill a sense of pride in the person accomplishing the work.

SUMMARY

Establishing a clear sense of direction about what a sponsor hopes to achieve in an assisted living project involves identifying common principles, agreeing on a common image and concept for the building, and establishing a therapeutic care philosophy. This chapter has discussed how to focus the design and evaluation process to achieve these ends. It has also reviewed important concepts and principles that help to clarify the meaning of assistive living as an alternative to institutionalization. Finally, it proposes nine criteria that can be used to classify and define assisted living environments.

Innovation and progress will continue to be made with regard to this housing type. It is important that our future attempts to improve assisted living address both issues of form and function. Architecture that is functionally correct only solves half the problem. The true power of architecture lies in its ability to delight, stimulate, and inspire residents, family members, and staff.

ENRICHMENT POSSIBILITIES

4

INTRODUCTION

Northern Europeans often use the word "possibilities" to describe the role architecture can play in stimulating choice and creative experimentation. Managers, architects, residents, and therapists often describe possibilities associated with room relationships, access to gardens, off-site views, equipment purchases, furniture placements, and unit features. For example, having food preparation and storage facilities in the dwelling unit provides residents with the possibility of exercising a range of independent behaviors. Exploring differing possibilities can expand the present and anticipated future needs of residents.

In reviewing the 100 European and 25 United States site-visited buildings, fifteen primary themes emerged. Each suggests possibilities that can add richness, depth of expression, program choices, and therapeutic stimulation to the environment. This chapter defines these themes, describes examples, and suggests ideas for future exploration. The following brief description of each theme is followed by a more detailed treatment.

15 THEMES

1. **Corridors as Streets**: Transforming the monotonous, dark, and depressing double-loaded enclosed corridors of the institution into an evocative, stimulating, light-filled walkway is at the heart of this theme. Enclosed streets in the Netherlands combine the charm and scale of the medieval street with the compactness and efficiency of the double-loaded corridor.

2. **Personalization at the Unit Edge**: Linking dwelling units to corridor spaces through windows and Dutch doors facilitates social connections between residents. The unit edge can be further personalized through accessories like street numbers, plants, paintings, photographs, doorbells, and light fixtures that give each unit individuality and definition.

3. **Atriums for Gardens and Activities**: Enclosed atrium spaces between rows of dwelling units can be used for lush, landscaped gardens, and/or engaging and stimulating activities. The atrium provides a protected and accessible area to pursue personal interests or to observe activities.

4. **Serving the Broader Community**: Housing projects for the frail often produce and deliver health and social services to vulnerable older people living in the neighborhood. Services are either provided to neighbors at the housing facility or delivered directly to the homes of nearby neighborhood residents. A service house with a restaurant that serves meals or a swimming facility that offers exercise therapy exemplify the former, while meals-on-wheels or a home-delivered health care program exemplify the latter. European housing for the frail serves both its immediate constituency and older neighborhood residents.

FIGURE 4.1 Many service houses are heavily community-oriented. The *Eskegården* service house in Beder, Denmark, delivers twenty times more meals for residents of the neighborhood through a meals-on-wheels program than for older residents of the center.

5. **Role of Landscape Architecture**: Visual tactile and olfactory contact with plant materials provides sensory stimulation that can enhance active physical therapy or support passive meditation. Because physically and mentally impaired residents have fewer opportunities to explore the neighborhood, the site often substitutes for that broader realm. Places to wander, to be active, and to disengage should be accommodated in outdoor landscaped rooms.

6. **Bending, Stretching, and Breaking Rules and Regulations**: Laws that regulate housing and services for the frail, as well as building code requirements, often narrow design choices and eliminate therapeutically beneficial activities. State regulations can limit the participation of older residents in self-maintaining activities of daily living, while building codes modeled after hospital rather than housing requirements often lead to highly institutional buildings.

7. **Options for Socialization and Observation**: The environment should be designed to facilitate opportunities for viewing on- and off-site activities. These include activities generated by residents, staff, visitors, and family members. The environment should enhance previewing, spontaneous social exchange, and opportunities for informal discussion because these activities stimulate residents and reduce depression.

8. **Therapeutic Design**: Mental and physical engagement are important aspects of therapy which can enhance competence and extend independence. Environments can support a range of therapeutic activities through the design of rooms and spaces that encourage physical and mental exercise. Intergenerational activities, pet therapy, and plant materials can play important roles in stimulating residents.

9. **Linking and Connecting Spaces**: Spatial variety as well as spatial relationships between corridors and common social spaces can stimulate opportunities for social exchange. They also provide variety and interest in the form of interior spaces of different volume, shape, and sequence.

10. **Inviting the Family**: Allowing families to more fully participate in caregiving partnerships with formal providers of housing and services is generally underexplored in housing options available in the United States. Making a building attractive to families for visits and overnight stays allows it to function like housing rather than a hospital, therefore encouraging family interaction. A major concern older people have regarding institutionalization is the extent to which it alienates them from family interactions.

11. **Realm Between Indoor and Outdoor Space**: Two fundamental spatial typologies exist between indoor and outdoor spaces at the edge of housing projects. The first type of space includes conditioned and protected interior rooms that overlook the surrounding site. These are manifest in the form of greenhouse enclosures, bay windows, window seats, and enclosed porches. The second type of space consists of outdoor areas which are defined and protected extensions of the building. These can include porches, arcades, lath houses, and balconies. Both types of space are useful to mentally and physically frail residents who are often not able to fully experience outdoor spaces.

12. **Dwelling Unit Features**: Less ambulatory residents often spend much of their time in their dwelling unit. It should therefore facilitate access, manipulation, safety, and convenience. Dwelling unit features designed to adapt to the changing needs of residents should be specified. A full kitchen and complete bathroom define a room as a dwelling unit. Privacy and the extra space available in a larger dwelling unit allow personalization and encourage family relationships.

13. **Residential Character and Imagery**: Ensuring that a facility appears as a residential rather than institutional building is a concern that goes beyond aesthetics. Residential references affect the attitudes and behaviors of residents, staff, and family, while institutional references alienate and isolate by discouraging family involvement and visitation. The appearance of the building, its organization, and the scale of individual rooms affect perceptions of residential design.

14. **Rooms with a Behavioral Purpose**: When spaces are designed without attention to purpose, function, and use, they often lack strong ideas about how to address therapy, engagement, activity, and socialization. When rooms are not well designed to accommodate reading, socializing, observing, card playing, and other small group activities, they can degenerate into undefined, unattractive lounges that are rarely used.

15. **Unit Clusters**: Clustering units in small groups encourages informal social exchange, helping behaviors and friendship formation. It also allows groups of dwelling units to be targeted toward groups of residents with specific needs and interests. Units configured around small lounges create physical configurations that add variety to the building's form and allow outdoor patio and garden spaces to play a more substantial role.

The following explores each of these fifteen possibility themes in more depth and detail by referencing examples from European and United States site visits that illustrate noteworthy solutions.

1. CORRIDORS AS STREETS

Introduction

The double-loaded corridor is an efficient housing circulation configuration that minimizes corridor length and maximizes density. However, the lack of natural light, the disorienting nature of the enclosure, and the lack of connection between the corridor and each dwelling unit create conditions that make it an unsatisfying architectural solution. Nevertheless, the compactness of the form has the potential to enrich places because it channels and focuses activity and movement. The best example of the potential for delight inherent in the density afforded by the double-loaded corridor is the richness of expression and focused activity of the narrow medieval street. Here, an intimate scale combines with activity, sunlight, and permeable window and door connections to make the street an engaging and exciting place.

FIGURE 4.2 The intimate physical dimensions of the medieval street are the inspiration for the Dutch enclosed street. In contrast to the typical double-loaded interior corridor, the scale, activity concentration, and available natural light make this a stimulating place for informal social exchange and the development of friendships.

Enclosed Street Schemes

Transforming the drabness of the double-loaded corridor into a delightful and animated streetscape is the architectural challenge. Fire code limitations designed to minimize the threat of fire and smoke make the connection between the corridor and the unit a stark impersonal linkage. Schemes that embrace the imagery and scale of the medieval street while protecting the corridor from inclement weather and allowing natural light to penetrate have the greatest potential for success. The enclosed streets of **Bergzicht**, in Breda, the Netherlands, **de Kortenaer**, in Helmond, the Netherlands, **Baunbo**, in Nr. Nebel, Denmark, and **Anholtskamp**, in Markelo, the Netherlands, pursue these ideas.

The Dutch *steunpunt* projects have taken this concept farthest. The idea behind the enclosed street is simply to provide protection for residents as they walk from their dwelling unit to adjacent common spaces, where activities are available to neighborhood residents as well as residents of the housing project (SEV, 1991b). In the Netherlands, severe west winds from the North Sea can be very unpleasant, especially in the winter when temperatures frequently drop below freezing. The enclosed sidewalk provides protection from the rain and snow as well as from winter winds. Because the enclosure is designed to shelter and not create an interior space, the character of the enclosed street seems related more to exterior environments. Plants are placed in pots and planters are trained against walls. Walls are brick and windows and doors open onto the corridor. The paving material provides the clearest reference to the idea of a street. In most instances it is made of concrete tile, occasionally banded with a darker color to create a pattern. The most striking feature, however, which differentiates it from an enclosed corridor is the quantity and quality of natural light. Sunlight penetrates even to lower floors, giving the place an outdoor feeling. Without sunlight, the corridor would appear conventional, not like an enclosed street.

The **Bergzicht** in Breda, the Netherlands (see case study in Chapter Five), is a three-story enclosed street, urban in form and character. The walkway on upper floors is attached to one wall and bridges over to unit clusters on the opposite side.

The **de Kortenaer** project in Helmond, the Netherlands, capitalizes on the linear nature of its site to create a long two-story corridor extrusion. The upper-floor walkway is placed in the center of the atrium, allowing natural light to reach the lower floor through light wells

FIGURE 4.3 The enclosed street at *de Kortenaer*, in Helmond, the Netherlands, is a two-story housing block which allows light to penetrate along two sides of a central bridge. Note the plants and the concrete floor pavers that give this space the character of an outdoor garden rather than an enclosed interior corridor.

on both sides. Light penetrates to the lower floor by bathing the walls of the lower floor, increasing the amount and consistency of natural daylight that enters the unit through windows that face the enclosed street.

The project is entered near the middle, where common spaces for residents and neighbors are located. The area above the common meeting room contains offices and storage. As the resident population ages, this space will be used to provide services.

The two edges of the project site are treated differently. Units on one side accommodate off-street parking and a walled garden that opens to the unit. The other side has a small private patio that opens onto a shared green lawn. Adjacent to the project entry door is an open stair that connects the two floors. A small enclosed garden and sitting area borders the front entry and community room. This area is popular with residents and visitors, who wait for the mail, watch others enter and exit, or pass the time of day by reading the paper in the filtered sunlight of the atrium. The plants located here give this space a gardenlike character.

The **Baunbo** project, in Nr. Nebel, Denmark, is a simple one-story enclosed street with a peaked roof framed in translucent plastic. The T-shaped configuration of the street corridor minimizes travel distance and adds to its interest. Plants in garden plots adjacent to entry doors and below unit windows reinforce the exterior garden character of the enclosed street. One unique quality of this setting is a lounge situated at one end of the corridor, which overlooks an adjacent main walking street in this small west Jutland town. The end of the street corridor has large doors, which accommodate vehicles when opened. Residents enjoy the friendly nature of this corner, because it allows them to watch street activity from a central place that is convenient to all residents. Each unit has a private garden, a popular feature with the primarily retired farmers who live in this rural township. Units in the 500–600-square-foot range are outfitted with full kitchens and have access to a range of health and social services.

Two-story schemes are the most visually complex and interesting projects. They are open enough to allow adequate amounts of natural light to penetrate the lower floor.

The Street as Metaphor

Utilizing the metaphor of the street as a connecting linkage between units and common spaces has interesting design possibilities. The **Skinnarvikens Servicehus**, in Stockholm, Sweden, has pursued this idea in a project on a heavily sloping site. Six compact point towers of six stories each house a total of 165 units. Each has a separate entry on the ground floor. However, a lower level links together all six buildings through an interior corridor system designed to resemble an exterior street. On this floor, the base of each concrete tower has a rough stucco finish that resembles the exterior treatment on the ground floor above. Streetlights, wood roofs above each tower's entry door, windows, and half walls create the illusion that the common rooms, doctor's office, laundry, multipurpose space, restaurant, and administrative spaces are parts of a small town conveniently linked together by a corridor. Skylights and one edge of the lower floor (open to daylight and view) are used to introduce natural light that furthers the illusion of the corridor as street.

FIGURE 4.4 The *WZV Anholtskamp* enclosed street, in Markelo, the Netherlands, clusters four units together. Eight units on each of two floors create a comfortable scale. Window and door connections to the street give each dwelling unit a sense of identity.

Summary

The idea of transforming corridors into streets that attract activity and link spaces together has powerful social implications. Streets in many cultures are active and engaging behavioral settings (Rudovsky, 1969; Gehl,1987). The enclosed street creates an attractive setting for social interaction. Furthermore, mitigating the influence of harsh weather makes the street usable year-round, and accommodating landscape materials softens it with a gardenlike character. Perhaps the most inspiring aspect of this idea is that it takes a problem space and turns it into an exciting and positive project feature, transforming a major liability into a noteworthy and unique asset.

2. PERSONALIZATION AT THE UNIT EDGE

Introduction

One of the most negative aspects of utilizing double-loaded, fire-rated internal corridors to connect dwelling units is the anonymity they exude, often rendering the hallway lifeless and boring. One way to counteract this problem is to design the unit entry to support activities, accommodate visual permeability, and utilize features associated with a traditional entry experience.

The unit entry can thus support ideas about individual expression, physical connection, and personal communication. When dwelling unit entry doors are clustered, they add spatial variety to the corridor and increase opportunities for informal social exchange. Furniture placed here can extend the personal domain of the unit into the corridor. Windows and doors can link the unit with the corridor, extending an invitation for neighborly exchange between adjacent dwellings.

In the **Captain Eldridge Congregate House**, in Hyannis, Massachusetts, a Dutch door, windowshade, and light fixture are used to signal a resident's desire for privacy or interest in conversation. Commonly understood conventions provide residents with effective ways of controlling the connection between their unit and semipublic corridor space.

In the **Vereniging Anders Wonen vor im Ouderen** project, in Breda, the Netherlands, a window that faces the corridor is used to signal neighbors about the availability of a resident for activities (SEV, 1991a). Depending on how a card is placed in the window, a desire to welcome guests or a resident's prolonged vacation can be communicated. Four separate messages are transmitted depending on the orientation and position of the card.

Sense of Individuality

Numerous elements can assist in creating a sense of individuality, including features that have their origin in the single-family house. Creating an alcove by recessing the door is one of the most common approaches. Symbolically this "protects" the visitor like the porch does on the outside of a house. Window and door openings can be used to transform walls from edges to seams. Windows can be controversial because they may reveal more of the unit interior to the passerby than might be desired and

FIGURE 4.5 The *Captain Eldridge Congregate House,* in Hyannis, Massachusetts, has created a porch alcove between a light-filled atrium and the unit edge. A double-hung window, Dutch door, light fixture, and windowshade allow residents to vary the openness of each unit kitchen to the semipublic domain of the project.

FIGURE 4.6 *Woodside Place,* in Oakmont, Pennsylvania, has used Dutch-inspired half doors to connect resident units to the corridor of each small bungalow house. This connection allows residents an informal but protected connection to the corridor. It also keeps wanderers from walking into other resident's rooms but still allows them to watch activities along the corridor.

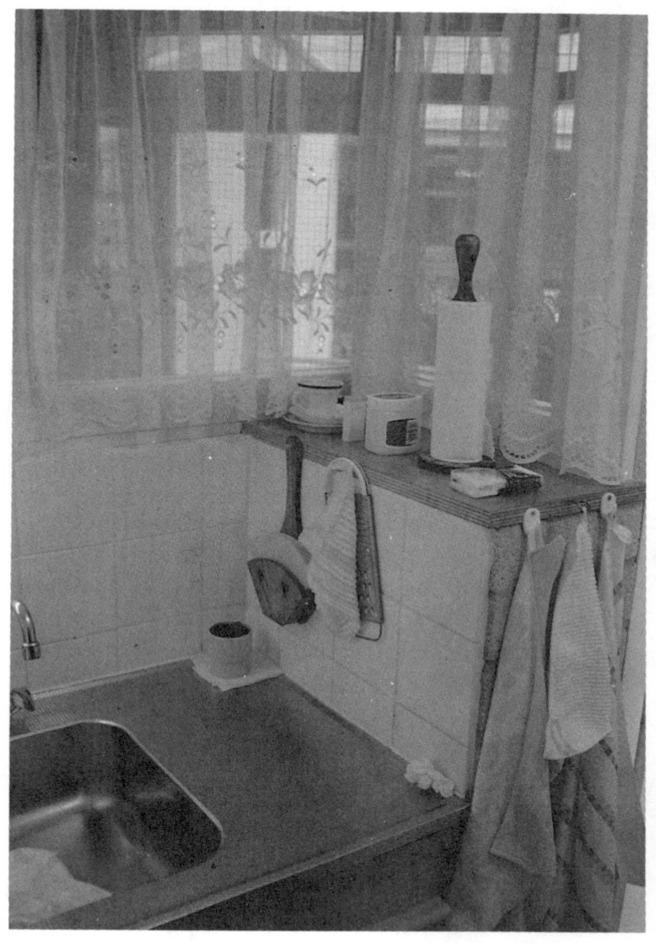

FIGURE 4.7 An interior window links the unit kitchens of the *de Overloop* project, in Almere-Haven, the Netherlands, to a double-loaded corridor, creating the opportunity to treat the corridor as a street. Strategically placed cabinets block the corridor view into the bedroom and living room.

because fire codes limit their size, configuration, and design. In many settings one-hour fire separations between corridors and dwelling units must be maintained. Windows, when used, are commonly located adjacent to the entry door and open to the kitchen.

The most common features used to define the entry are a mailbox or mail slot, a shelf for resting packages or displaying items, a doorbell or door knocker, a street number, a wall-mounted light, a peephole for viewing out, a welcome mat at the foot of the door, and a planter for landscape materials.

Dutch projects seem to utilize these features more than any other culture. More than three quarters of the projects visited in the Netherlands employ either windows, alcoves, Dutch doors, or cluster plans to create a sense of personal expression at the unit edge. Over a third of the Dutch projects visited utilized half doors (Dutch doors) to open units to the corridor while simultaneously protecting residents' privacy and security.

Personalizing the Corridor

Another important aspect of the unit entry is its ability to communicate, through symbols and artifacts, personal aspects of the resident. This adds variety and interest to the corridor while giving visitors an impression of the interests, abilities, and experiences of the older resident living on the other side of the door.

In most United States projects, corridor walls are treated as if they belong to management. They are often decorated with photos, accessories, and artwork selected by management rather than by residents. One of the most interesting aspects of Dutch housing for the elderly is the personalizing of building corridors with resident photographs and artwork. In these projects, items displayed in corridors adjacent to unit entries are personal, including family photographs, drawings, and paintings that had been created by residents. Seasonal decorations and events such as birthday parties or grandchildren's visits also provide an excuse for a temporary display near the entry to share the event with others.

Further, the corridor can be treated as a semiprivate space extension of the unit. In single-loaded corridors with views to atriums and off-site activities, resident furniture placed in the corridor transforms this space into a semiprivate unit extension where plants, a table, and reading material can be kept. Thus, the corridor connects unit entries through a series of individual semiprivate extensions of each unit.

Finally, use of plants can also be an inviting way to personalize the corridor at the unit entry. For this to be successful, enough light must be available for plants to flourish. Skylights located above entries, exterior window walls adjacent to single-loaded corridors, and enclosed street atriums provide this possibility.

Summary

Personalizing the unit entry through windows, doors, and accessories creates a seam rather than a barrier between the sanctity of the unit and the semipublic nature of the corridor. Artwork, photos, and plants on corridor walls allow this space to become an extension of the unit. Rather than settling for corridors that are lifeless and anonymous, they can become places that reveal to visitors the values and symbolic associations that have meaning to older residents.

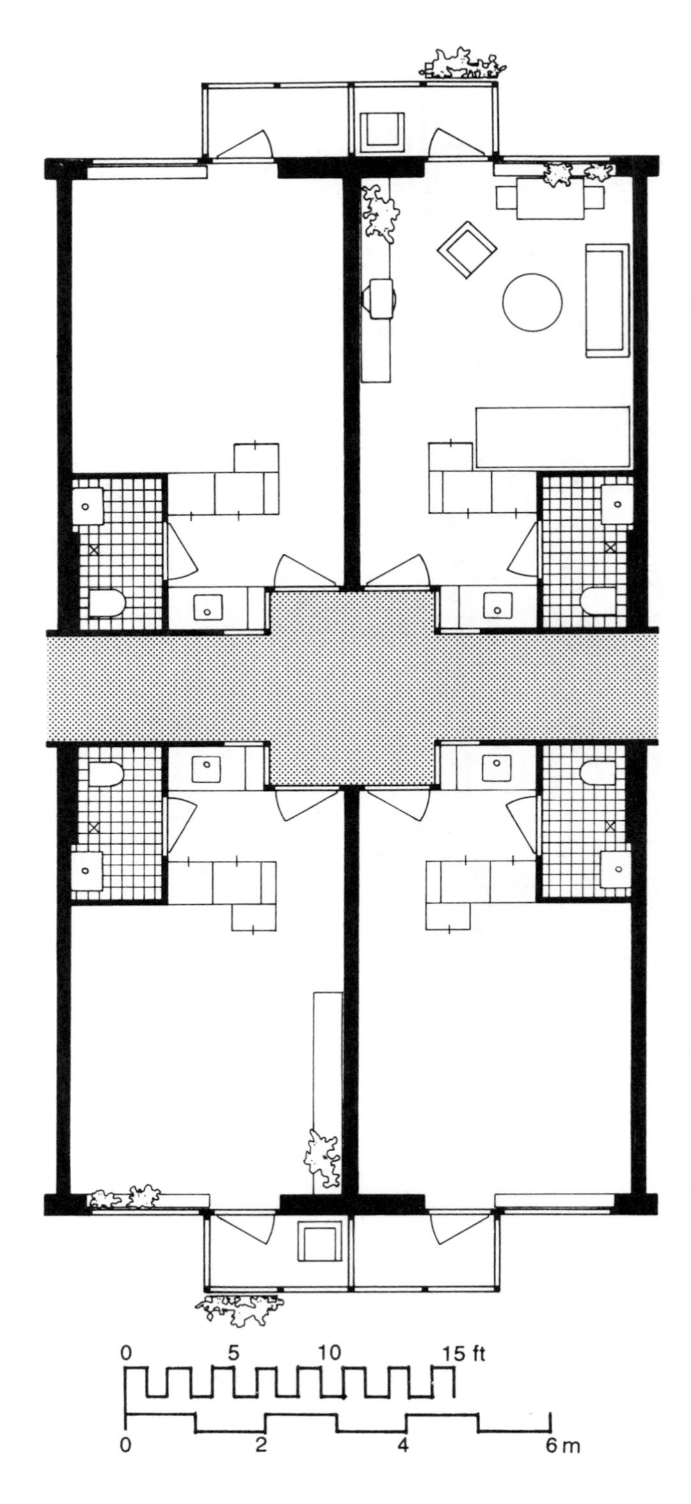

FIGURE 4.8 Dwelling units in double-loaded corridor portions of the *de Overloop,* in Almere-Haven, the Netherlands, are clustered in groups of four. Dutch doors, corner windows, and a widened corridor alcove allow the unit and corridor to have a house and street-type relationship. Small kitchens and balconies give each compact unit a sense of completeness and variety.

3. ATRIUMS FOR GARDENS AND ACTIVITIES

Introduction

Creating an atrium involves enclosing the area between dwelling units with a transparent or translucent envelope that shelters and secures the space. Atriums can be treated as formal interior spaces, which are conditioned, ventilated, and heated to a standard equal to that of other habitable interior rooms, or they can be minimally treated by a protective enclosure with efforts taken to safeguard against temperature extremes. In other cases, atriums can be outfitted with mechanized sliding skylights that open the atrium to the sky when weather conditions permit.

The area created by this enclosure can become a focal point to support common activities, or a garden setting for passive pursuits like reading or observing. The character of the space and the purposes it serves are also influenced by the use of landscape materials and the amount of control available over the temperature and humidity of the space.

FIGURE 4.9 The atrium at the *Palvelutalo Esikko* in Turku, Finland, has a green wall of plants that gives it character and warmth. The atrium floor is constructed over a swimming pool and thus cannot support a garden. Modern pendant light fixtures suspended from cables tethered to the roof define sitting areas in the atrium.

Special Character of Space

In most cases the atrium offers a relatively large protected space that introduces sunlight into the center of the building, providing a place for individual and group activities. Atrium spaces appear to be particularly good for older people whose patterns of neighborhood use are limited because of mental and physical declines or concerns about safety and security. Creating visual and physical access to a garden area for retreat, a community room for socializing, or a perch for observing activities gives residents an additional highly accessible protected space.

Nineteen of the European projects contained atriums. Six of these were designed as gardens with soil floors, with the remainder designed as hard-surface activity spaces. Many of these activity spaces, because of the natural light available, utilize extensive potted plants to add to the garden character of the space. The most intriguing spaces were those that captured part of the atrium for garden use while allowing the remainder to be used for informal or group social activities.

Atrium Applications

The enclosed multilevel streets described earlier differ from most atriums in that they rely on a single corridor to access units on two sides of the street. Most atrium projects contain two parallel but separate single-loaded balcony corridors that service units separated by a multistory light well.

FIGURE 4.10 The garden atrium at *Moerwijk,* in The Hague, the Netherlands, is located at one end of a double-loaded corridor. A Y-shaped corridor configuration is terminated by a sloping roof of glass suspended from a space frame. The garden visible from the project's front entry opens to a conveniently placed community room.

However, atriums are also created from Y-shaped diverging corridors. The **Moerwijk** project, in The Hague, the Netherlands, takes advantage of this corridor configuration to create a sloped glass pyramidal enclosure. The tropical nature of the plants inside make it a special place to visit in the winter. A combination of paved and soft surface floor treatments allow it to serve a variety of purposes, including small group social meetings as well as community events like a regularly scheduled flea market. The **Kruistraat** project, in Utrecht, the Netherlands, is another Y-shaped building. Here a double-loaded corridor branches into two single-loaded balcony corridors around a triangular atrium, with a wall of glass overlooking an exterior garden space at one end. This is particularly effective in providing a sense of orientation.

Although most atriums are linked to residential units, some projects utilize the atrium as the activity center of the project, surrounding it with common activity spaces. **De Drie Hoven**, in Amsterdam, **de Overloop** in Almere-Haven, the Netherlands and **Degneparken**, in Dianalund, Denmark, use this approach, giving the atrium a communal civic character (Residenza per anziani a Almere-Haven, 1984; Auf dem Polder, 1985).

Noise has been a problem in the **Humanitas** atrium in Hengelo, the Netherlands, because of hard wall and floor surface specifications. The cooperative house **Vereniging Anders**, in Breda, clustered fourteen units around a common stair and elevator connecting all three floors. Residents often meet one another when leaving or returning to their unit. Atriums in the **Lundagården** project in Lundsbrunn, Sweden (Lundagårdens Sjukhem, Lundsbrunn, 1984), **de Kiekendief**, in Almere-Stad, the Netherlands, and **Dr. W. Drees**, in The Hague, are designed to be opened to the sky through a mechanized tract that slides the roof canopy to one side.

First-Floor Atrium Uses

The **Jan van der Ploeg** and **Nybodergaården** projects, in Rotterdam and Copenhagen, respectively, have specially configured balconies that allow residents to sit on the edge of the atrium and watch activities below.

First-floor areas of atrium projects are assigned a number of different functions. In the **Nybodergaården** project the space is used as a bar, a café, and for general socializing. In the **Dr. W. Drees**, in The Hague, and the **Degneparken** service house, in Dianalund, Denmark, billiards, card playing, weaving, and library spaces are defined through furniture placements. In the **Viherkoti** project, in Espoo, Finland, a portion of the atrium is used as an open physical therapy and exercise area. **De Kiekendief**'s atrium, in Almere-Stad, the Netherlands, is a large space that contains several kiosk-style buildings within the atrium for food service and groceries.

Overheating appears to be a problem in some atrium projects designed with noninsulated roof materials when mechanical or natural ventilation is the only form of air-conditioning. Most projects use thermostatically controlled

FIGURE 4.11 The atrium garden at the *Rosenborg Centret,* in Copenhagen, daylights interior spaces while creating an attractive place to enjoy morning and afternoon coffee. A pyramid-shaped glass roof tops the atrium. Office spaces for workers on one side have openable casement windows that overlook the interior garden. This allows these office spaces to have natural light and a view.

FIGURE 4.12 The two-story atrium in the *Degneparken* service house, in the small Danish town of Dianalund, has defined several activity areas through the placement of cabinets, plant materials, and furniture. This is an active and vital space which is filled with light and activity most of the day. A two-story housing block adjacent to the atrium overlooks the space and visually connects residents with the activities here.

dampers that open automatically when temperatures reach a critical level, or if smoke from a fire is detected. Insulated translucent panels are being installed at one project that had problems with overheating. In another project, air-conditioning is used for one month in the summer when it is considered necessary.

One of the most intriguing atrium examples found in **Carlslund Park**, in Upplands Väsby, Sweden, consists of a small U-shaped single-story building surrounding a shared enclosed atrium that accommodates six units. The intimate scale of the grouping and the nature of the protected atrium space encourages residents to extend activities into this semipublic domain.

Summary

Atrium spaces serve a number of important functions in housing for the frail elderly. They introduce natural light into the center of projects and allow landscape materials to flourish in central common areas and gardens. They support a range of resident behaviors from meditation to group activities. Finally, they allow the older frail with ambulatory difficulties to have convenient access to a protected space, which has many of the characteristics of an outdoor garden.

4. SERVING THE BROADER COMMUNITY

Introduction

United States facilities that care for older frail people are often cognizant of the opportunity to serve older frail people living in the surrounding neighborhood, but are rarely organized to pursue this idea. These neighborhood residents often require either home care services of a more sporadic nature or monitoring services like emergency response. Housing projects that deliver community services to neighbors along with servicing residents create a hybrid model, which combines the continuity of care associated with assisted living, the service support system of a senior community center, and the outreach efforts of a home care agency. Combining these three housing and service components allows one setting to respond to a broad array of community needs. This is the foundation for European housing and service systems for the frail. Three main approaches to providing community service occur in Europe.

One: The community-based center offers a range of services at a central site that neighborhood residents can visit. This can include services as diverse as occupational therapy, physical therapy, meals, adult day care, respite care, health care checkups, recreational opportunities, and social activities.

Two: The second approach involves the production of services at the center which are delivered to neighborhood residents in their own homes. Home-delivered meals, home help, home health care, emergency medical response, and transportation are a few of the most common.

Three: In the third, housing is co-located with other compatible land uses. Child-care services, outpatient clinics, pharmacies, restaurants, and grocery stores are typical choices. These mixed-use models have the potential to connect housing with a range of compatible land uses, thereby further integrating the project into the surrounding context.

Mixed-use housing makes a further contribution to the resources of the city. The desire to serve a community population gives each project a strong civic presence in the city. The scale of a typical forty- to sixty-unit service house with community spaces is large enough to have an impact on the character of the surrounding neighborhood.

Service Model of Community Support

In Europe, projects are sized for the service needs of older frail residents in an immediate one- to two-mile radius as well as providing purpose-built housing for the frail. These numbers achieve the economies of scale necessary for efficient service production. If the community outreach component was not included, projects would need a considerably greater number of units. Looking at the system this way allows us to imagine community supports, coordinated from a centralized location that provides help to people living in adjacent neighborhood housing or in service flats within the same building. For each older person, services are custom-fitted to needs and are consistent with the desire to stimulate the highest level of self-maintenance.

Just as hospitals have shifted from continuous hospital bed monitoring to a system of outpatient care, so will the housing project of the future need to think of its role as a community resource and not just an institution focused on the needs of a few select in-patients.

The European building type that holds the greatest promise for advancing knowledge about the integration of neighborhood-based housing and community services is the service house (Bull and Lise-Saglie, 1991). This concept was very popular, appearing in all five northern European countries and represented by 44 out of the 100 site-visited buildings. The variation in models and scale was great, the smallest model consisting of seven units and the largest, over 200 units. Projects averaged forty to sixty units in size. Rural and small-town settings often had comprehensive models that borrowed program elements from hospitals. The service house in a small town might substitute for a rehabilitation hospital and a nursing home. In larger cities, service houses were present in distinct neighborhoods and were assigned responsibility for serving the needs of local constituents. These projects contained lively, heavily patronized community centers, where meals were provided to neighbors as well as residents. The number and type of service components varied for each setting. Some were self-contained, while others were connected to health facilities.

Mixed-use models located above or adjacent to shopping centers or town squares are sited overlooking streets and sidewalks that accommodate access to retail services. These locations are popular for passive observation by residents. How a project fits, complements, and improves the urban context is another important aspect of how it serves the broader community.

FIGURE 4.13 Home care nurses at neighborhood-based service houses take care of older people in the surrounding neighborhood who need assistance. The density of development in Danish towns and the difficulty in finding parking often make it easier to ride a bicycle to the homes of neighborhood residents than drive a car.

Restaurant

Nearly all service houses contain a restaurant that serves residents, as well as older people from the surrounding community. It is generally the most popular and heavily used space. In a majority of settings the restaurant attracts a group from the neighborhood that is younger, more mobile, and more socially outgoing than the resident population. Frail residents generally living in upper-floor dwelling units have the option of taking meals at the center or having them delivered to their unit.

In projects like **de Klinker** and **de Gooyer**, in Amsterdam, meals are taken in a first-floor community room that is used for socializing, card playing, and other group activities later in the day. In these two service houses, a bar located between the restaurant and the project's front entry creates a pre- and post-dining area for socializing and activities like billiards. Many such settings were designed to resemble, in form, appearance, and function, the traditional Dutch neighborhood "brown café." One project reported popularity with such a range of neighborhood age groups that management, in an effort to optimize use by older residents, resorted to asking clients their age before serving them, to screen out the under-sixty-five crowd. In these similar settings, beer and stronger drinks are available for purchase at slightly below retail prices.

Recreational Activities and Meeting Rooms

Some service houses are designed to function more as a community center than others. In the **Midtløkken** service house, in Tønsberg, Norway, community volunteer groups have offices and meeting space at the center. At **Anholtskamp**, in Markelo, the Netherlands, community meeting rooms are available to local organizations, including service clubs and the Boy Scouts.

Some projects like **Forsmannsenteret**, in Sandefjord, Norway (Forsmannsenteret, 1989), are so popular and heavily used by neighborhood residents that residents of the housing feel they have little claim to community spaces. Given the amount of interaction and the potential for friction that exists between residents and neighborhood "outsiders," it is remarkable so little conflict actually occurs. One reason could be that residents accept the model, which implicitly requires the sharing of facilities with neighbors. They realize that to support a wide range of services and activities, fixed costs must be distributed over a larger constituency. For example, the swimming pool in the **Midtløkken** service house couldn't have been constructed and maintained with the revenues generated by the fifty-seven residents of the project alone.

Occupational and Physical Therapy

Service houses contain a variety of equipment and personnel devoted to building and maintaining the general health and fitness of residents and community members. Physical and occupational therapy (ergotherapy) are common. The Danish service house, however, appears to be the most sophisticated employer of therapeutic space, equipment, and personnel. The Danish commitment to rehabilitation is serious and focused. Although other countries recognize the desire to include rehabilitational services, rarely are they as well outfitted or staffed.

In the **Rygardscentret**, in Hellerup, Denmark (Rygardcentret i Gentofte, 1981), the occupational and physical therapy rooms are situated around outdoor courtyards and are used for a range of therapies in the summer. Wood, brick, concrete, and grass surfaces are used to exercise different leg and foot muscles through walking therapies.

Health Care, Hospice, and Rehabilitative Services

Service houses in smaller towns typically broaden their community role to include more in the way of long-term care and rehabilitative services. The **Degneparken** service house, in Dianalund, Denmark, is a good example. Located in a small town on the west edge of the island of Sealand, the service house has a six-bed infirmary that serves as a transition ward for residents moving from the local hospital back to the community. It has a complete training kitchen and access to sophisticated therapy equipment.

Hospice arrangements are also common at many service houses. Cases are typically handled on an individual basis, but residents are rarely moved to a nursing home. Most service houses have the capability of increasing nursing service intensity to a twenty-four-hour schedule. For most hospice patients, the progression of disease often provides a reasonably predictable time frame for death. Families often operate as care-giving partners with formal care professionals in these arrangements.

The **de Gooyer** service house, in Amsterdam, has a population that has aged in place since the building opened eleven years ago. Here, a special training program has been developed in conjunction with a nearby nursing home. Nursing home professionals train service house workers to deal with difficult cases and are used as problem-solving consultants. A number of interesting experimental programs in the Netherlands explore hybrid building types that combine health care-dependent individuals with the frail in need of personal care. These experiments are designed to test the limits of residential programs. Most service houses report a remarkable record in retaining residents and minimizing transfers to nursing homes. In many cases the only residents who cannot be managed are dementia residents who can't be deterred from wandering out of the building or hurting themselves. Most service houses employ a broad range of strategies to help residents age in place.

Day-Care and Respite Services

Elderly day-care programs in many Dutch and Danish projects serve community residents who need intellectual and physical stimulation within a monitored group setting. Day-care programs like the one at **de Overloop**, in Almere-Haven, the Netherlands, are designed to accommodate both mentally frail residents and community members. Day-care members normally take their meals together in a small group but are isolated from the mentally alert resident and community populations. One unusual program at the large **de Drie Hoven** center, in Amsterdam, offers night care service for families with mentally impaired members who are unusually active at night.

Swimming Pools and Saunas

Although widely recognized for their therapeutic exercise value, swimming pools are often restricted from small projects by their cost of construction and operation. However, in about a dozen service houses, this feature has been developed jointly with local government. In the **Palvelutalo Esikko** project, in Turku, Finland, a swimming pool located under the central atrium was financed jointly by the local municipality and the housing project. It is offered on a time share basis to various community groups. In the **Midtløkken** project, in Tønsberg, Norway, and the **Gulkrögcentret**, in Vieje, Denmark, large swimming pools with therapy programs are provided as part of the service center's commitment to the health care and exercise needs of the surrounding community.

In the **Kotikallio** service house, in Helsinki, a swimming pool and sauna are shared with a rehabilitation hospital located on top of the service house building in a mixed-use configuration. The hillside site allows on-grade access for the rehabilitation hospital on the fourth floor and the swimming pool on the lowest floor.

Home Helpers and Nurses

Most service houses contain offices, meeting space, and a supply depot for district home help and nursing personnel who meet daily to organize their work schedule. Thus, residents and older people in the surrounding neighborhood rely on home service personnel whose base of operation is in the service house. Organized this way, home helpers and nurses are in a position to monitor needs and to suggest when vulnerable community residents should consider moving to the service house. The central location for the district also stimulates better communication between nurses and home helpers. Care teams are generally organized as small groups but often operate within the same service house.

Occasionally, mixed-use projects combine outpatient health care with the service house. The **Opmaat** project, in Monster, the Netherlands, has interconnected these two building types. In the **Brahenpuiston Asuintalo**, in Helsinki, an outpatient health center is part of a mixed-use complex centered on an urban plaza in front of the building. In such models, district doctors, who oversee the health care needs of the elderly, young children, and families, operate from a location adjacent to the service house.

FIGURE 4.14 This swimming pool located partially below grade, at the *Martensund Servicebus,* in Lund, Sweden, overlooks the surrounding landscape and has access to natural light. Both of these qualities make the pool a delightful place to exercise. As in most service house arrangements, the pool is shared with older residents of the surrounding neighborhood.

Community-Delivered Services

In addition to providing a range of social, therapeutic, and health care services, service houses also coordinate home health and home help. Meals-on-wheels is one of the most common services. In projects like the **Eskagården** service house, in Beder, Denmark, the number of meals delivered to the community is twenty times greater than the number of meals prepared for residents in the project.

Another important service provided to residents living in the community is emergency medical care. In Boras, Sweden, the emergency response office in the local service house monitors residents as well as hundreds of older community members.

Many centers have developed programs that help community residents with home improvements. The **Flesseman Center** and **de Klinker** projects, in Amsterdam, (Wooncentrum Nieuwmarkt Te Amsterdam, 1990), have home repair programs that assist residents of the surrounding neighborhood to secure their housing against crime and increase its physical accessibility.

Mixed-Use Strategies

Fitting housing and service buildings into an urban community can be accomplished through a mixed land-use development strategy. The **de Gooyer** project, in Amsterdam, has faced several retail spaces along an important shopping street adjacent to the first floor of the service house building. These lively land uses allow the block-long service house to add life and activity to the street. A restaurant tenant regularly fills the sidewalk in front with tables and chairs, adding interest to the street.

Children's Day Care

A day-care center for children on the protected inner courtyard side of the **Brahenpiuston Asuntalo,** in Helsinki, provides an animated playground for residents to watch. Children's day care is a relatively common addition to Swedish service houses for the elderly. At the **Aspens** project, in Linköping, a children's day-care center was developed by remodeling several unpopular grade-level housing units. Children take their lunch in the service house restaurant along with older residents. In the **Vanhainkoti** home for the aged, in Pori, Finland (Vanhakoiviston Vanhainkoti, 1989), the site is shared with a children's day-care center. The library in the elderly center is outfitted with small chairs and tables for children who are involved in joint activities with older residents.

Sharing Community Spaces

Sharing space with other age groups is of increasing interest to municipalities looking for ways to balance growing public service budgets. In the **Skelagar** project, in Århus, Denmark, housing is located adjacent to a community center that offers a variety of special use spaces for the elderly as well as the general community. The location, next to a shopping center, provides residents with good accessibility to retail goods and services.

One of the most interesting mixed-use projects is **T-1**, in Linköping, Sweden. Located on a recently abandoned military base, this new community includes family housing, retail stores, schools, and two service houses. Spaces programmed for the school and service house are combined. A gymnasium, sewing room, workshop, and classroom located on the first floor of the elderly service house are shared with local elementary school students and families living in the surrounding neighborhood. Sharing these spaces adds to the age-integrated character of the service house and reduces the duplication that would occur if separate exercise and workshop facilities were provided for the school, the community, and the elderly.

Urban Design Strategies

Projects can contribute to the design of the city in numerous positive and beneficial ways. One of the most interesting things for residents to watch is the movement of traffic and pedestrians. Many projects overlook streets, allowing residents a picture window view of the action outside. The restaurant at the **Flesseman Center**, in Amsterdam, overlooks the busy Nieuwmarkt Square subway entrance. At the **Old People's Home and Health Center**, in Oitti, Finland, a pedestrian walkway that passes through this building at a lower level connects two adjacent areas of the town. A decentralized dining room overlooks this curved pathway, which is filled with bicycle riders and pedestrians most of the day.

Other projects are conscious of their responsibility to preserve and enhance the city. At the **Aspens** project in Linköping (Altenheim Aspen in Linköping, 1980), a popular pedestrian street on the edge of the development has maintained the old artists' studios located here. The **Gulkrögcentret**, in Vieje, Denmark, is located between the central train station and a major hospital. A plaza and sidewalk system, designed to accommodate pedestrian flow through the project, can be viewed by residents from several interior vantage points.

Other projects have recognized their relationship to the neighborhood by using the building's design to connect residents to nearby retail goods and services. The **Midtløkken** project, in Tønsberg, Norway, and the **Apian Palvelukeskus** project, in Valkeakoski, Finland (Apian Palvelutalo, 1988), face urban streets on one side and are open to plaza and park spaces on the opposite side. In both cases, U-shaped building configurations create courtyards at the rear of the project that are linked to the city through adjacent plazas and parks. These direct connections encourage residents to use them, infusing the courtyards with vitality and activity.

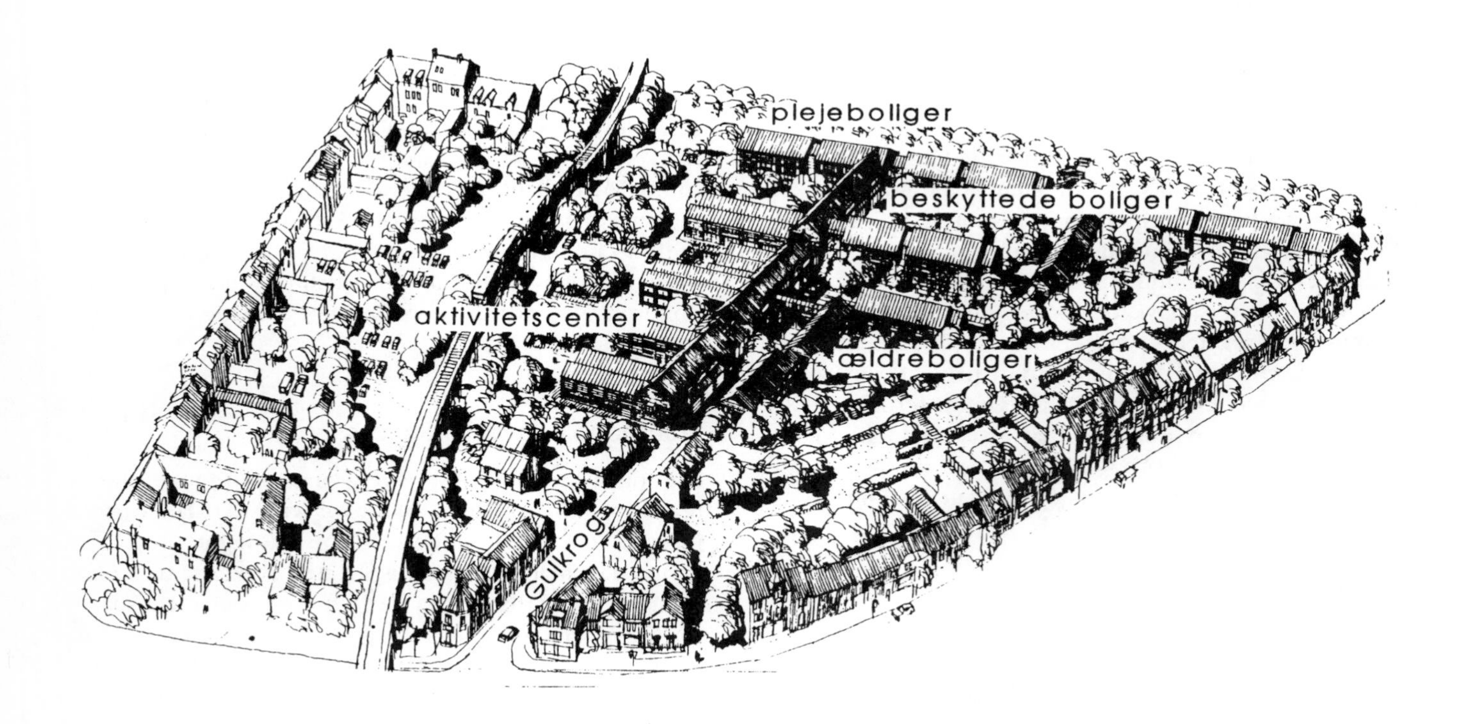

FIGURE 4.15 *Gulkrögcentret* was designed as an urban renewal housing project near the central city of Vieje, Denmark. Housing was patterned after adjacent buildings so the project would fit neatly into the neighborhood. A service center, nursing home, and supportive housing are connected by a network of courtyards and pedestrian streets.

FIGURE 4.16 *Gulkrögcentret* lies between a major hospital and the train depot. The building layout accommodates workers who walk through the project to the train station. Resident social areas were placed in locations that overlook heavily traveled pedestrian paths.

Summary

European attitudes about connecting projects to the community both physically and operationally are far superior to attitudes in the United States. We have developed a system of private facilities that often turn their back on the community, ignoring its presence. We rarely design senior centers in conjunction with housing, and conversely rarely use publicly financed housing projects with large, spacious community and activity spaces to serve neighborhood residents. Senior centers for years have focused on the recreational and social needs of residents at the expense of helping the older frail stay independent in the community. Rarely does a senior center provide outpatient health services or act as a coordination point for home health and homemaker personnel. We have developed a fragmented system that rarely coordinates housing with services. Northern Europeans have a tested model in the service house, which is well worth exploring and interpreting in the United States. Only a few projects like **Lincolnia Center,** in Fairfax, Virginia, have experimented by co-locating adult day care and assisted living with a senior center.

5. ROLE OF LANDSCAPE ARCHITECTURE

Introduction

The full appreciation of plant materials as aesthetic, utilitarian, therapeutic, and sensory design elements is the key to understanding their potential for influencing the site appearance and residents' quality of life. In both European and United States facilities, plants regularly appear as therapeutic elements. Greenhouse and solarium spaces are frequently programmed into assisted living settings, and residents will often nurture plants in their own rooms that are gifts or have been brought from a previous residence. Exterior landscapes have the potential to provide spaces for contemplation and retreat, as well as places for exercise and sensory stimulation.

Plants can trigger memories of salient places and events or they can be experienced as passive elements in a controlled view situation. They can also be used to attract birds, butterflies, and other wildlife, or to animate the environment through sounds and smells. Wandering gardens have been developed for dementia residents and can symbolize places and spaces that are familiar and comfortable. Rose and herb gardens contain plants with pungent, sweet, or savory aromas that can stimulate senses in a way that is pleasant and memory-provoking.

Finally, the seasonal and diurnal nature of landscape materials allow us to mark time and create associations with important seasonal occasions and special past events. Given that older people spend considerable time in their dwelling units and that assisted living residents are often physically impaired, these qualities of the natural environment may play an even more important role. The landscape as a potential canvas for design expression provides a number of possible avenues for adding stimulation, complexity, and variety to a setting.

Landscape Strategies

The design of housing should be approached with the premise that landscape architecture plays an important role by defining outdoor "rooms" and relationships to the surrounding neighborhood. With regard to site treatments, projects often fall into one of two categories. Buildings are either placed in the middle of a parklike setting, or fit within an urban context that defines outdoor gardens through courtyards.

Building in the Park

De Drie Hoven, in Amsterdam, is an example of a housing and nursing complex designed within a parklike setting. An earthen berm at the perimeter of the site isolates it from its suburban context and creates a "green edge" from the outside and for residents inside. Rainwater moving through the site to a retention basin flows through plazas and adjacent walkways. A blanket of trees planted in the parking lots disguises cars and blacktop with a leafy green umbrella.

Several projects located next to city parks take advantage of adjacency by linking up with them. The **Strandlund** project, in Charlottenlund, Denmark, has clustered housing around pedestrian and vehicular streets as fingers within a large park. These streets concentrate activity and movement, allowing the remainder of the outdoor space to be free and open. Turf block, which accommodates automobile and pedestrian access, creates a soft, parklike quality to the avenues. The relatively small number of less frequently driven cars allows these semipaved areas to maintain a soft green appearance.

Courtyard Forms

The **Tornhuset** project, in Göteborg, Sweden, is located in an older industrial district. A new, relatively stark six-story U-shaped building shelters a garden courtyard adjacent to an early twentieth-century former factory administrative building that contains a picturesque clock tower. The simple courtyard configuration is open on one side to accommodate views of an adjacent waterway and to connect the garden with a park located along the waterway. A bridge over the canal allows residents to walk to the city center.

Many Dutch projects understand the beauty of water as a landscape element. Several projects like **Moerwijk**, in The Hague, and **de Overloop**, in Almere-Haven, have bodies of water bordering the building that create an attractive landscape view from common interior spaces.

Each country has made unique additions to the vocabulary of ideas about landscape design. The Dutch *höfje,* a seventeenth-century courtyard housing type, was originally designed as social housing for widows and unmarried women of limited financial means. **Höfje van Staats**, in Haarlem, clusters twenty units around a 100-foot by 140-foot courtyard. The paved perimeter of the

FIGURE 4.17 The *Flesseman Center*, in Amsterdam, has unit projections that facilitate street watching. The center courtyard of the project, however, is designed as a heavily landscaped visual oasis for social interaction and meditation.

courtyard provides a place to sit in the sun and an area to nurture potted plants. A single large tree forms a focal element. The courtyard is located behind a large retail space, which completes the street edge and isolates the project from street noise. Residents enjoy the convenient location, solitude, and security of the courtyard, which adds to the project's social life, allowing friendships to flourish. The original housing was constructed in 1730. A recent remodeling added inclined elevator stairlifts to some units to help older disabled residents continue living there.

In Norway, the *tun* is an outdoor courtyard space originally used to contain and protect animals in rural eighteenth-century farmsteads (Hauglid, 1989). Today the tun is being interpreted in cluster housing schemes to create meaningful places from the spaces left over between buildings. **Raufosstun**, in Raufoss, Norway (Raufosstun, 1979), has created back porches that link dwelling units that surround an outdoor courtyard focused on a centrally located stand of birch trees. Tuns were created by building clusters originally placed on the land to complement topography and capture views. These seemingly random clusters rarely rely on an orthogonal grid. The resulting geometry of the protected inner courtyards adds variety and sense of fit with the surrounding natural topography.

Sensory Stimulation and Therapeutic Intervention

One promising aspect of outdoor spaces is the possibility they provide for sensory stimulation and therapeutic application. They can complement exercise programs or provide a place of seclusion for private conversations with family members. They can support active uses like picnic spaces or croquet, or provide places to passively watch birds, butterflies, and small mammals. They can engage residents in personal activities like nurturing plants or provide a serene aesthetic backdrop for events like family photographs. Some projects have used plantings to aid wayfinding by locating specimens that vary in size, configuration, and color.

Woodside Place, in Oakmont, Pennsylvania, gave each of its three courtyards a unique orientation. Although each is the same size, the layout of the sidewalk and the location and type of plants differ for each. One courtyard is designed with plants of enhanced color, another is situated to attract birds and animals, and the third has specified savory- and sweet-smelling plants. With the right collection of plant materials, it is not uncommon to regularly attract as many as twenty-five different bird species. Even more species can be attracted in the fall and spring during migration.

At the **Jewish Home for the Aged**, in Reseda, California, jasmine has been placed in planter beds of differing heights in a small garden that accommodates wheelchair users, residents who are seated, and those bending over to smell flowers. The design involves a cluster of intimate spaces for contemplation and small group discussions. The Danes have utilized outdoor areas in nursing homes quite differently than what is normally seen in the United States. Gardens or paved spaces connected to each resident's room are common in most Danish projects.

A Swedish nursing home, **Vickelbygården**, in Skärblacka, inspired by this Danish tradition, has created

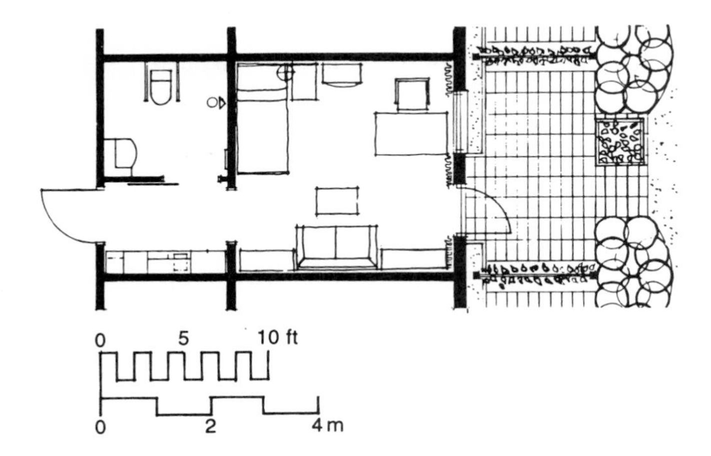

FIGURE 4.18 The *Vickelbygården* nursing home, in Skärblacka, Sweden, is designed after a Danish model that emphasizes autonomy by specifying a small kitchen and an accessible outdoor patio. The compact unit is single-occupied and contains a large size bathroom for transfers. Note the use of a sliding barn-style door to the bathroom. Even though this unit is 275 square feet it seems bigger because of the visual connection to the outdoors. (Source: Christofferson and Ruhnbro, 1987).

both private patios and shared courtyards for displaying colorful plants. The courtyards, surrounded by colorful plantings in raised beds, are easy to access from inside and are visible from the dining room. Each residential unit opens onto a small paved patio with a single twenty-four-inch-square raised wooden planter box. The planter boxes become a visible and attractive way to add color, texture, and individuality to each patio.

In the United States projects, Alzheimer's wandering gardens have also become quite popular. At the **Motion Picture Country Home and Hospital**, in Woodland Hills, California, a wandering garden with a looped pathway connects several different plant ecologies. One area outfitted with an aviary, park bench, and water fountain symbolizes a civic park. Another location along the walkway contains a swing and a place to be alone. A third portion of the garden with a slight slope has a "babbling brook" flanked by pine trees, which resembles the ecology of the foothills. The looped sidewalk takes residents by each of these locations, providing a range of different areas.

The **Anton Pieckhöfje**, in Haarlem, has clustered the entrances to six group homes for dementia victims around an enclosed sidewalk and an open garden. The protected sidewalk allows residents to use it at night and during inclement weather. Alcoves located along the walkway overlook the courtyard. Chairs and tables provided here allow residents to rest or converse with family members, other residents, and staff.

Swedish group homes for dementia victims have conceptualized outdoor areas for sitting, socializing, picnicking, and gardening. The **Lönngården** group home, in Nässjö, Sweden, has a vegetable garden adjacent to several fruit trees, where residents grow food they eventually harvest and consume.

FIGURE 4.19 The *Vickelbygården* nursing home features a patio outside each dwelling unit. Each patio contains a small raised planter, which residents and their families can use to cultivate flowers during the summer.

FIGURE 4.20 The courtyard of the *Kvarteret Karl XI* service house, in Halmstad, Sweden, contains a fragment of the old medieval city wall. It was retained and restored as a feature of the courtyard. The courtyard landscape pattern is derived from the location of battlements.

Historic Reference

Historic reference is perhaps easier to accomplish in cultures where the history of civilization goes back a thousand years or more. In excavating the site for the **Kvarteret Karl XI** service house, in Halmstad, Sweden, a portion of the old medieval city wall was uncovered. Fragments of the wall run diagonally across the courtyard and have been used to create sitting areas in the courtyard.

The **Munke Moss** project, in Odense, Denmark, was designed to replace an older home for the aged. A fountain, salvaged from the old building, was placed in the center of the courtyard, and designed as a meditation space to remind residents, visitors, and staff of the history of the place.

The entry court of the **Vickelbygården** nursing home, in Skarblacka, Sweden, was sited around a large old oak tree. Because of the tree's beauty and symbolism, it was preserved and the building designed around it. When the tree eventually died, a local artist was commissioned to carve the trunk into a sculpture as a tribute to the memory of the tree instead of removing it. Now everyone who visits the project sees a wood sculpture prominently displayed at the front of the building.

Summary

Landscape architecture plays an important role in enhancing therapy, stimulating the senses, encouraging social interaction, and ensuring privacy. For the older frail person with limited mobility, a well-landscaped site can be an important source of stimulation, substituting for contact with the larger neighborhood. Using the landscape to its fullest extent should be based on an understanding of how it can enhance the quality of each resident's life. These ways can be varied and individual, including active engagement such as gardens, picnicking, and exercising or passive observation of objects and activities on-site.

6. BENDING, STRETCHING, AND BREAKING RULES AND REGULATIONS

Introduction

Rules and regulations are established to guide practice, ensure safety, and provide a reasonable level of quality assurance. In theory, policies and regulations have their place in a system that takes responsibility for an at-risk population. Sometimes, professional advocacy in the form of stringent, uncompromising standards is necessary. However, the design of environments for older frail people in the United States is a casebook example of how commonly held good intentions have been transformed into a nightmare of overlapping, contradictory, and counterintuitive constraints. Rules often result in environments that are safe but sterile, and caregiving philosophies that are hamstrung by inappropriate emphasis on accountability.

In an attempt to establish policies that safeguard older people, we have greatly narrowed the range of possible creative solutions. Regulations that are prescriptive establish standards in perpetuity, which are at best tolerable, and at worst an affront to the dignity of residents and the competence of staff. Rules that establish corridor widths and travel distance to nurses' stations ignore new technologies and caregiving strategies. In doing so, they create deeper ruts in a flawed system of care management that desperately needs reform. No one feels good about placement in a nursing home. Hospitals are far more acceptable to a broader range of people than nursing home settings. Yet our system of public policies, regulatory constraints, and code requirements push us in the direction of continuing to advocate errors and mistakes well recognized in past evaluations.

Although the nursing home is in great need of reform (Hiatt, 1991), there are only a handful of experimental models exploring ideas that challenge antiquated rules and question inflexible guidelines. United States society has proliferated defective models of long-term care by ignoring opportunities to test new ideas.

One way to understand alternative approaches is to inspect renegade United States models that avoid rigid regulatory definitions. Occasionally, these models are the result of enlightened public policy, such as the assisted living experiments in Oregon. More often, however, they are housing arrangements that squeak by regulations in one state but are considered too radical to be sanctioned in others. The **Rosewood Estate** model, in Roseville, Minnesota, described in Chapter Five, is one such example.

Regulations often affect both the physical and operational environments. The costs of outmoded, inefficient, archaic, and rigid requirements can be measured in the waste of resources devoted to accountability rather than caregiving and in human costs measured by the constraints placed on the autonomy, privacy, and dignity of residents. With financial resources dwindling, and the demand for present and future housing for the frail increasing, we must begin to employ innovative and constructive reforms quickly.

Innovative foreign models of housing for the frail can provide a unique lens through which our assumptions and current practice can be questioned. These housing models, designed for people with the same mental and physical impairments as those in the United States, are different from a physical, spiritual, and operational perspective. Looking at these examples improves our understanding of how the physical environment can liberate or incarcerate. Rules and laws have been instituted to overcome management and caregiving problems with the best of intentions, but often with major unanticipated negative side effects.

Danish Plegehem

Each of the five European countries visited has developed unique, innovative models of long-term care. The liberal sharing of information between countries with related but different cultures has allowed ideas to be transplanted, refined, and improved. One of the most intriguing models is the Danish *plegehem* (pronounced play em), equivalent to the United States skilled nursing facility. Residents, although similar in average age, appear more active and ambulatory than their United States counterparts. Making comparisons is always difficult, but the Danish plegehem offers some interesting differences.

FIGURE 4.21 The Danish nurses' station is designed around the idea of promoting conversations between residents and staff. Instead of locating desks behind high counters, tables are used to promote staff conversations with residents.

1. Decentralized Organization

Most facilities are organized around ten- to fifteen-person clusters. Each group eats in small, separate dining rooms. Each cluster is independently managed and administered by staff that often employ residents on management committees. Physical environments conform in shape and scale to these clusters. Multistory models can have fifteen residents to a floor while others may contain courtyards designed around clusters of ten units.

2. Small-Scale Facility

Although a few facilities were large and included several different levels of care, most were in the forty- to sixty-unit range. The most efficient cluster of residents is not driven by licensing requirements as it is in the United States. The forty- to sixty-unit size is pleasant in scale because it is small enough to fit unobtrusively in the neighborhood, and conforms in appearance to other forms of housing.

3. Nurses' Stations

In the United States, regulations regarding access to the nurses' station establish a rigid geometry that limits the possibilities for the building's form and organization. In Danish nursing homes, the rules are different. Distance from the nurses' station to a resident's room is not a primary criteria for organization. The outcome is a philosophy about caregiving that appears much more humane. Models are not driven by recordkeeping for the sake of keeping records. Caregivers are more concerned with keeping residents fit and independent than with conforming to rules and accountability standards. Thus, more creativity and professionalism exist in decisions about therapy and the benefits of managing the greater risks, which result from an emphasis on independence and self-responsibility.

4. Corridors

Single-loaded corridor schemes are common and dwelling units are often organized around decentralized lounges and nurses' stations. While double-loaded corridors are used sparingly, courtyard and atrium forms are common. The major advantage of single-loaded corridor and courtyard forms is the amount of daylight that reaches common rooms and connecting spaces. Longer distances are overcome by scooters that enable nurses to get around more quickly. The decentralized clustering of units and the local management of these clusters also reduces travel distances, making these schemes convenient and effective.

5. Resident Rooms

Privacy in the form of single-occupied units is a strongly held norm. Standards dictate private single rooms of 225–250 square feet for each resident, which includes a room and a private toilet. Two-bed rooms are generally only available for couples.

6. Lighthearted Convivial Environments

The most curious aspect of the Danish plegehem is the commitment to an open, highly participatory philosophy of management, while maintaining a rigorous commitment to a program of physical and ergotherapy. Numerous facilities feature bars that sell alcohol in the front lobby. These are frequented by residents, visitors, and family, who visit and take part in the convivial and lighthearted atmosphere it creates.

7. Commitment to Therapy

On the other hand, physical and occupational therapy are viewed as a necessary lifestyle dimension of the daily life of each resident and everyone participates to the best of their abilities. The emphasis on wellness and challenge has been accepted because residents and staff generally understand the consequences of sedentary disengagement. The philosophy of these facilities has adapted the structure and discipline of regular therapy and the informality and conviviality of social communication and interaction.

Swedish Group Homes

Thirteen group homes for dementia victims were visited in the five northern European countries. The largest concentration of group homes is in Sweden, where the movement to promote this housing alternative has been steadily growing over the last seven years (Kuller, 1991; Almburg and Paulsson, 1991). These small-scale housing environments are structured around a simple and compelling idea about care. Homes of six to eight residents are accommodated in remodeled social housing, or new purpose-built arrangements. Three site-visited service houses had group homes adjacent or connected to their facilities.

The story told by Annelie Hollo of the Swedish Planning and Rationalization Institute (SPRI) is one that gives reason to pause and consider the American situation. When the Swedish group home movement began, there was a great deal of concern over its "informal residential appearance" and more relaxed style of caregiving. Physicians and nurses protested the movement, arguing that psychogeriatric institutions were better suited to the needs and behaviors of demented older people. Studies were structured to compare behavioral outcomes between patients living in a conventional psychogeriatric ward and those living in a hospital ward that had been remodeled to resemble a group home.

Controlled Study of Group Home

The studies conducted at Varnhem Hospital, in Malmo, Sweden, took place over an eight-month period and involved a simple intervention. A dining room was decorated in a residential style as it might have been in the 1930s and 1940s. Instead of eating off plastic trays, patients had porcelain dishes and served themselves. The staff also ate with patients. The outcomes were rather amazing given the minor adjustment to policies and the environment. Residents were happier, more susceptible to social contacts, and their dietary intake increased.

Followup studies have been conducted with control groups testing the social adjustments that patients have experienced when they have moved out of a psychogeriatric hospital into a smaller, more intimate group home. Preliminary outcomes show a significantly higher level of social interest and social activity for the intervention group than the controls. Other studies continue to be pursued that are examining instrumental activities such as aggression, incontinence, and behavioral outbursts.

Another startling aspect of the group home movement is the fact that costs are regularly 20–30 percent less than placement in a psychogeriatric hospital. The humanity of the situation was increased and the costs were reduced. As might be expected, the group home movement has since received support from politicians and local municipalities and now has little opposition. Several monographs are now available as documentation of the experiences of group home facilities throughout Sweden (Beck-Friis, 1988; Almberg and Paulsson, 1988; Malmberg and Oremark, 1991; SPRI, 1990). Four qualities of these facilities deserve attention and further discussion.

1. Small Group Size

Most homes are organized around households of six to eight residents. This contributes to an intimate residential feeling and accommodates caregiving ideas such as encouraging residents and staff to take meals together around a single dining room table. The residential scale of the setting and its "family" feeling becomes compromised if this number is exceeded. Group homes have often been paired or clustered to reduce the cost of management and oversight, especially during nightshifts.

2. Residential Form

Projects have taken on a range of different configurations. Initial projects were located in remodeled social housing complexes, sometimes on the upper floors of high-rise buildings. Later projects employed the imagery of single-family houses in detached dwellings. Older,

FIGURE 4.22 The group home movement for dementia victims in Sweden centers around the therapeutic philosophy of normal daily activities. Residents are encouraged to help prepare meals, set the table, bus dishes, or help serve dinner. Laundry, gardening, and trips to the grocery store are also used as shared therapeutic activities.

larger single-family houses and apartments have also been remodeled for this purpose. However, accessibility problems caused by narrow doors and stairs have limited this approach. Some of the most interesting experiences involve mixing group homes with other family housing arrangements. The **Hasselknuten** project, in Stenungsund, Sweden, reviewed in Chapter Five, combines a group home with thirteen family housing units in a two-story L-shaped building. The **Anton Pieckhöfje**, in Haarlem, accommodates a cluster of six group homes on the first floor while utilizing the second floor for fourteen units of family housing.

3. Normal Daily Activities as Therapy

The group home is founded on the philosophy of engaging residents in normal activities of daily living. Residents are involved in meal preparation, encouraged to set the table, help serve food, bus dishes, and participate in supervised food preparation tasks such as baking bread and preparing vegetables. Laundry is another activity that involves residents in folding and ironing clothes, and placing clothing in the washing machine and clothes dryer. Many homes creatively pursue a range of other ideas.

Some take residents on walks in the neighborhood or to nearby parks. In other settings residents accompany staff to the store to purchase food and help carry it home. In some facilities, gardens are tended in summer and harvested in fall. In one project, two residents were observed returning with an aide after fishing at a nearby lake in the morning. The fish were cleaned and eaten by staff and residents for lunch that day. Passive pursuits include reading the newspaper, listening to music, watching television, and completing crossword puzzles. Using regular daily activities as a form of therapy allows the home to appear natural.

4. Personalized Contact

Every resident is assigned a contact person, who learns the background of that resident, knows their family, understands their preferences, can administer and organize their medications, anticipates their limits, and provides information to others about their participation in household activities. The key to the therapeutic potential of the group home model is the contact person. This person pays careful attention to tasks residents can master, activities from which they can gain satisfaction, and is in a position to encourage family participation in structured ways. The group home is a personal, relatively unrestricted, and normalized model of care. Residents rarely move from a group home to a psychogeriatric facility or nursing home. Group homes with stairs and accessibility constraints can sometimes pose a problem for residents with wheelchairs. However, purpose-built arrangements are adaptable environments that allow residents with physical and mental impairments to age in place and die in the facility.

The struggle for acceptance of the group home movement offers a lesson that can be applied to institutional reform in the United States. Testing the model through a demonstration project under controlled circumstances allows a clear assessment of its benefits over the status quo. This type of approach should be instituted in measuring the benefits of noninstitutional group living arrangements in the United States.

Local Nursing Homes

The local nursing home is a Swedish experiment centered around the idea of providing skilled nursing care to small decentralized clusters of residents (SPRI, 1979). Begun in the early 1980s, this movement was never fully realized because of changes in government policy that moved away from nursing homes toward a policy of expanded home care. Rather than designing institutions to resemble group home clusters, local government has encouraged older people to stay in their own homes as long as possible. The reason for this is twofold. Providing home care is often less expensive than institutionalization and older people prefer to stay at home as long as possible. When they can no longer cope with expanded service needs at home, then placement in a group home or service house is sought. In Sweden, as elsewhere in Scandinavia, a greater proportion of the older population than ever before receives all service needs at home.

The **Solgård** nursing home, in Tranemo, Sweden, exemplifies the local nursing home concept well. This thirty-two-resident home clusters four residents around each of eight living rooms. Two living room clusters share a family style dining room where staff and residents take meals together. This building form reflects a philosophy of care that is both personal and informal. Visual and physical access to the outdoors is enhanced along with the amount of natural light. It is a pity so few of these model facilities were built to replace older homes. Those that were constructed are far superior to the old-style conventional nursing home.

FIGURE 4.23 The *Solgård* local nursing home, at Tranemo, Sweden, uses decentralization to break down the scale of the nursing home. Four units are clustered around a living room. A convenient centrally located dining room serves eight units. Spaces flow into one another, avoiding the need for long corridors. The thirty-two units at Solgård reportedly operate at the same level of economic efficiency as a nursing home four times as large.

Dutch Experiments

The Dutch have focused their efforts on testing a number of innovative housing ideas through the construction of demonstration projects and service programs. Stuurgroep Experimenten Volkshuisvesting (SEV) is perhaps the best-known sponsoring and monitoring group. It often co-ventures demonstration programs with local housing and health care organizations. SEV projects are testing a range of new ideas including:

1. The viability of communal and self-help housing arrangements,
2. The combination of nursing care with traditional personal care in residential settings,
3. The creation of outreach strategies disseminated from service houses that support older community residents in conventional rental housing, and
4. The remodeling of older institutional facilities with small rooms into buildings with larger units that emphasize independence.

The concerns of efficiency and cost containment are evident here just as they are in all countries. Some experiments involve the suspension of existing regulations and policies, while others only tinker with the rules. The freedom to test ideas that involve combining housing and healthcare is clearly worth pursuing in the United States.

In **de Westerweeren** project, in Bergambacht, the Netherlands, nursing home patients are mixed with traditional home for aged residents. The more independent setting that results offers all residents options regarding food preparation.

Design and Organization of the Physical Environment

One of the most interesting physical design attributes of European facilities involves the use of atrium buildings. Compared with the United States, European code compliance standards for fire suppression equipment and mitigations seem relaxed. Projects are rarely fire-sprinklered. Instead, most contain elaborate systems for smoke detection and exhaust. European building materials rely on brick and concrete more than in the United States, where wood construction is more common. However, even concrete and steel buildings in the United States have fire safety measures that by comparison appear exaggerated.

While American hotels and hospitals can afford to install expensive fire suppression systems that comply with these standards, nursing homes cannot afford the cost of sophisticated devices required to comply with atrium-type open spaces. In the United States, fundamen-

FIGURE 4.24 The Dutch and Danes have designed many of their long-term care facilities around light filled single-loaded corridors and courtyards. The *Dr. Drees* project in The Hague, the Netherlands, uses single-loaded corridors and a large atrium. Neither country requires nurses' stations to be a "set distance" from patient rooms. The outcome of this policy has been very satisfactory in terms of operational and aesthetic impacts.

tal systems of fire control are installed but exciting sophisticated designs of linked and overlapped spaces are rejected. Residents are safe from fire and smoke in nursing homes, but they are denied the delight and interest of a variety of room sizes and heights. The resulting effect is a collection of dull fire-safe corridors that lead to safe self-contained rooms, much like a rat maze.

Another trade-off made in United States facilities is the design of fire stairs with limited glass exposures that restrict the amount of natural light entering a building. European facilities seem universally committed to using fire stairs as a source of borrowed light. The delight of the compact Hertzberger plan at **de Overloop**, in Almere-Haven, the Netherlands, is realized in part from the transparent stair enclosures that flood the interior with light. In the **Viherkoti** facility, in Espoo, Finland, a central atrium with windows that open to common spaces on the upper floors is bordered by a glass stair enclosure that faces south, allowing a generous amount of light to enter the core of the building. Furthermore, windows that face into the atrium are openable, allowing ventilation and spatial connections between the atrium and upper-floor corridors.

European Code Considerations

Emphasis on controlling noise also seems to be more of a concern in European projects that face busy streets or have access to common corridors. In Finland, double doors are often employed in these situations, with double windows used to insulate units from sound and harsh winter temperature differences. In Sweden, **Stocksundstorp**, a new project designed to overlook an adjacent freeway in Stockholm, has added enclosures to all balconies facing this noise source.

A further code difference from the United States system is the relaxed European attitude toward keeping corridors free of moveable furnishings that constrain the width to less than eight feet in nursing homes. The **Vastersol** project, in Jönköping, Sweden, is a local nursing home adaptively remodeled from an older acute-care hospital. The double-loaded corridor scheme contains plants, antique furniture, and chairs. Concerns about impeded access were considered overly cautious since effective fire suppression strategies in the past thirty years have involved isolating residents in their room when a fire occurs. Why would using the hallway to exit bedfast patients be encouraged when that would be far more dangerous? The same reasoning could apply to United States facilities with perhaps the same conclusion.

At the **Nybodergaården** facility, in Copenhagen, the **Kvarteret Karl XI** service house, in Halmstad, Sweden (Servicehus Halmstad, 1986), and the **Jan van der Ploeg** project, in Rotterdam, wide corridors accommodate chairs, tables, umbrella tables, books, and plants. Clear access is five to six feet in width. Are these projects foolhardy in overlooking concerns regarding exiting, or do they take a careful look at the benefits and costs of the setting and choose to err on the side of making the settings more humane although slightly more dangerous? One aspect of viewing solutions from a different cultural perspective is that it reveals our own subjective biases and values.

Finally, materials with high friction coefficients are used throughout European nursing homes. Materials like brick, stone, roughsawn wood, tile, and rough-textured stucco are common in corridors and meeting rooms. In the United States, nursing home codes limit the placement of abrasive materials, with the resulting outcome of overreliance on hard plastic laminates and paint. These materials give corridors a plain bland look while inadvertently creating flammability problems. Brick in the **Dronning Anne-Marie** nursing home corridors, in Copenhagen, has been laid to expose its interior cells to absorb sound. Additionally, the rough texture of the brick helps to deflect and break up sound. The design of care settings in Europe is clearly less constrained by needless regulation.

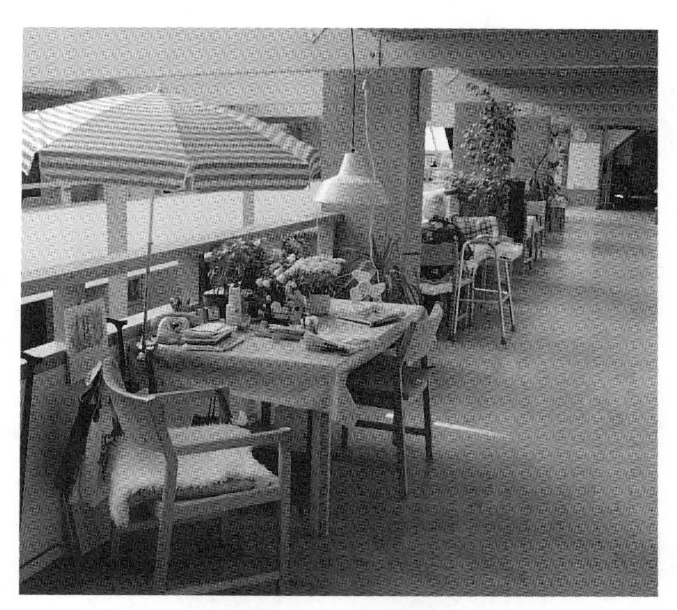

FIGURE 4.25 Wide single-loaded corridors at the *Nybodergaården* nursing home, in Copenhagen, are used by residents as an extension of their units. This space overlooks a busy light-filled atrium. Residents have outfitted this space with furniture. Here, a resident has placed a beach umbrella to protect the space from the afternoon sun.

United States Circumstances

In the United States, local fire officials trained as fire fighters often review drawings for complicity with regulations. Because each local fire fighting unit has limited equipment and experience, local control is commonplace. Fire fighters have a vested interest in making the setting safer rather than more attractive or humane. They haven't been sensitized to understand this problem from the caregiver's perspective. In Europe, fire compliance officials are better able to view these issues from a broad perspective. Social policies that set standards for a humane lifestyle are more widely held. Fire officials may also feel less responsibility to overprotect because they have faith in the system and its ability to optimize quality of life for older people.

In the United States, the frustration with deplorable conditions in some nursing homes has led to rigid physical standards for many aspects of the environment. It is far easier to specify construction materials and fire safety requirements than it is to control the treatment received by older residents or the potential abuse they may suffer. In limiting the use of residential materials, we have added to the inhumane and institutional appearance of the environment which also hardens the attitudes of staff about the nature of the setting.

Management Reform

Some of the most amazing contrasts between European and United States settings became apparent in discussions with European care providers about ways they involve older residents in the day-to-day life of their residence. At the **Dronning Anne-Marie** centret, small group clusters of resident rooms are organized into separate administrative units with a great amount of autonomy. Decisions about day-to-day operations are based on the philosophy of each unit, which employs residents in establishing policy. Bi-monthly meetings involving staff and residents are held to discuss problems and establish new initiatives. Furthermore, hiring at the ward level is done jointly by the administrative team and the residents living there, utilizing a committee comprised of 50 percent staff and 50 percent residents. Staff members claim that older residents are often the best judges of promising personnel, asking penetrating and revealing questions that professional staff are too timid to raise.

Another interesting aspect of care delivery in Denmark is the way in which information about resident well-being is communicated to other professionals. Care diaries record the daily experiences of residents in an effort to communicate this with other shift workers and health care professionals. Medical records and traditional

FIGURE 4.26 Nursing homes in Denmark have sought to decentralize units into small clusters of twelve to fifteen residents. At the *Dronning Anne-Marie* centret, in Frederiksberg, residents and staff have regular meetings to discuss hiring decisions and management policies.

charts are used to document medications and vital signs. The diaries are very unique, often involving comments patients make about their own perceptions of health status and wellness. Sometimes residents write in the diaries, sometimes they dictate how they feel to a health care worker, and other times their nurse writes a professional opinion. The diary becomes a communication device rather than a way to control information about the resident, since the emphasis is clearly on care provision rather than accountability.

Nurses' stations are not used to separate patients from the staff because the job of "reporting" is downplayed in contrast to time spent directly with patients. Danish nurses found it impossible to believe that alcoves with high counters are created for American nurses to work behind. The Danish impression is that the job of reporting on patient status is part of the nursing process and is therefore best carried out by interacting with patients rather than being isolated from them.

Another European policy that seems beyond the restrictive policies of most United States facilities concerns residents taking vacations away from the facility. In Sweden and Denmark, this remarkable custom involves taking residents, including the most debilitated, by bus and train to Nordic summer houses, French hotels, and Spanish beaches. This is done with the agreement of the older person and their family and is often considered one of the most important therapeutic opportunities of the year. Given our litigious system, imagine how impossible this practice would be in a United States nursing home.

Care Philosophy of Self-Management

A philosophy found throughout the Netherlands, which has strong support in other countries, is an emphasis on self-management. The Dutch call this nursing approach "care by sitting on your hands" or "putting your hands in your pocket." This philosophy of care emphasizes coaxing residents to do as much as they can for themselves. Traditional nursing practice, by contrast, is viewed as taking away responsibility rather than enhancing mastery and self-esteem.

The integration of the family in an assessment model of care planning is the basis for **Rosewood Estate**, in Roseville, Minnesota, and the programs of **Rackleff House**, in Canby, Oregon. These United States projects have an approach to care management that encourages self-care, interdependence, and family involvement. The analogy can be made to a relay race. Often family caregivers are running a marathon like it is a relay race. The responsibility of the older person, like the baton, is handed off when the caregiver becomes exhausted. The process should be a cooperative one, more like a three-legged race than a relay, where the responsibility for the older person is a partnership between the family and the formal caregiver. Regulations, liability concerns, and limited past experience with these ideas have created roadblocks that frustrate our vision and keep us from involving the family more directly in the caregiving process.

Summary

The philosophies described above capture the general thrust of the Scandinavian approach to providing care for the frail elderly. What is important to remember is that many ideas involving self-management, participatory governance, and therapy through normal daily activities run counter to laws and regulations initiated in the United States to safeguard the elderly. In United States nursing homes, when small-motor activity therapy like separating peas from their pods is employed, the food is normally thrown out rather than eaten because it violates health standards. Nursing homes can be cited for allowing residents to deliver in-house mail or to raise and lower the flag.

When resident gardens are turned under, as they are in many states because the regulations interpret this as a form of involuntary servitude, the system has overstepped its boundaries of safeguarding residents. Taking away activities that bring pride, a sense of satisfaction, and a feeling of usefulness to residents is wrong. Our European neighbors have managed to create a long-term care system that is not needlessly hamstrung by pointless and arbitrary rules that destroy opportunities for self-initiated resident activity and emphasize staff accountability over caregiving. We have much to do in addressing the diverse needs of older people who must be understood as unique individuals and treated as such, rather than contained in lifeless, boring warehouses.

7. OPTIONS FOR SOCIALIZATION AND OBSERVATION

Introduction

The mentally and physically impaired are often unable to use neighborhood services and may find that constraints brought on by increased frailty keep them from leaving the security and safety of the housing environment. Under these circumstances, it is important to recognize the need to think of the building as a microcosm of the city. It should include perches, corners, niches, retreats, alcoves, porches, and small-scaled rooms that offer opportunities to socialize with other residents, staff, and family members. Such places also allow residents to observe the activities and movement patterns of staff, other residents, visitors, and family members. These two interrelated but different activities of observation and social exchange must be accommodated in spaces that are appropriately designed and located. The most popular and heavily used social areas make it easy for residents to join and leave the group, and are situated in places with views of on- and off-site activities.

Views off-site into the adjoining neighborhood or adjacent streetscape are very popular. Viewing off-site activity can occur from a common space within the building or from an outdoor vantage point. Sitting areas with good views that are elevated from the street provide additional perceived protection from unpredictable or threatening street activity. Major sources of interesting visual diversity include activities created within the facility, residents and visitors entering and exiting the site, outdoor recreation opportunities, and small mammals. Even letter carriers and delivery trucks can provoke curiosity. Finally, the building itself can generate circulation patterns that are fascinating to watch.

Pedestrian circulation and staff work-movement patterns are a rich source of visual diversion for residents. Nurses' stations and front entry lounges are often documented as the most active gathering places for residents of nursing homes, because of the potential for observing activity (Koncelik, 1976).

Previewing

Creating places where residents can gather for spontaneous social exchange or for scheduled social and recreational activities is necessary for a successful, well-formulated spatial layout. The work of Howell (1980) and Zeisel (1981) suggests how influential circulation patterns and preview opportunities can be in stimulating social engagement. Previewing can be a catalyst that transforms observational activity into direct social exchange. The social dynamics of living in a group setting create additional complications that make previewing even more important in this context. Some residents enjoy encountering one another while others do not. A successful setting allows residents to control visual and social contact.

Several buildings are worthy of study with regard to basic issues of previewing. The **Captain Eldridge Congregate House**, in Hyannis, Massachusetts (reviewed in Chapter Five), is designed with several locations on the upper floor and stair landing for viewing the dining room and lounge space below. Residents can wait for the elevator or rest on the stair landing and preview social spaces on lower floors before entering them. In the **Degneparken** project, in Dianalund, Denmark, a similar opportunity is provided. A large two-story central space is the location for group activities, weaving, and reading. Residents who live in dwelling units that flank the space on the upper floor can see who is in the space before entering it.

Another effective way to allow previewing and entice residents into participating in activities is to make activities visible from along corridors. In the **Kvarteret Karl XI** service house, in Halmstad, Sweden, the first-floor corridor is designed to encourage people to walk through the building to a bus stop on the opposite side. Exterior light fixtures and finishes are used along this interior to give it a "street" character. Various activity spaces are arrayed along the corridor that links resident units with the restaurant and common activities. Residents can look into activity spaces through large glass wall panels that effectively allow previewing. They also introduce residents and visitors to the potential for involvement in a range of different activities such as ceramics, weaving, sewing, and art. Several benefits are associated with visually linking activities to the walkway. *First*, the route is made more interesting by a diversity of activities. *Second*, the visibility of group activities gives the building an active and engaged character for residents and visitors, demonstrating the spirit of the place in a powerful way. *Finally*, residents and neighborhood visitors can window-shop for activities without obligating themselves. Previewing in this context plays an important role in the success of the place.

Observation Opportunities

An important initial design study should be an assessment of the vitality and interest of the surrounding context. This will identify potential locations for views that can add to the popularity of sitting areas. Numerous urban projects directly engage street life through views from dining rooms, decentralized lounges, or dwelling units. The **de Gooyer**, **de Klinker**, and **Flesseman** projects, in Amsterdam, are located adjacent to streets that are lively. In the **Baunbo** project in Nr. Nebel, Denmark, residents meet around a table for morning coffee and watch children walking to school and young people bicycling to work. The view creates a catalyst that establishes the basis for social activity.

The **Runby Service Center**, in Upplands Väsby, Sweden, is located above the community library and overlooks a neighborhood shopping center. A small restaurant near the entry overlooks a busy walkway and a large plaza where neighborhood children play. In other projects, adjacent playground areas for children have created the focal point for observation activities.

Active, animated views of street life are also important and interesting to residents. A Boston housing project described in a study of outdoor spaces use (Regnier, 1985) illustrates this point well. An architect with a site adjacent to a busy street chose to orient the building away from the noise, clutter, and pollution of the street and toward a passive, quiet, and serene courtyard. This site planning decision resulted in a strange pattern of use. Every afternoon, residents carrying light aluminum chairs bypassed the serene tranquility of the courtyard and placed their chairs along the noisy, gritty street edge, where visual excitement abounds. Although the building was purposely designed to avoid this exposure, residents went out of their way to experience it.

On-Site Activity

There are many activities associated with a project that provides services. Meals are prepared, delivery trucks unload, and nurses and home helpers come and go. Capturing views of these activity patterns is also important. The nexus of this activity is the front door. Most projects take advantage of this by creating places near the entry for residents to watch outside activity and interact with one another.

At the **Strandlund** project, in Charlottenlund, Denmark (Strandlund, en boligbebyggelse for aeldre i Charlottenlund, 1983), a bar and fireplace located adjacent to the entry define a small area where four to five tables are located with excellent views of the pedestrian sidewalk, the street, and the entry path to the service house. Other entry lounges like the **Forsmannsenteret** project, in Sandefjord, Norway, use a fireplace, bookcases, a greenhouse enclosure, and booth seating to add individuality and interest to the entry lounge. In the **Kuuselan Palvelukoti** project, in Tampere, Finland, a platform adjacent to the front entry overlooks the garden atrium. Sunlight streams into this portion of the project during the day. Residents can enjoy their morning coffee here, watching people enter and exit the building, or take in

FIGURE 4.27 Many service houses like the *de Gooyer* project, in Amsterdam, are designed to reinforce lively city land-use patterns. In this project, the majority of the first floor fronts a market street and has been devoted to mixed-use retail. This makes the street edge more vital and gives residents a convenient place to shop.

FIGURE 4.28 Activity space walls at the *Kvarteret Karl XI* service house are completely transparent. Residents and visitors can see into each space and preview activities before entering. The glass case in this photograph is designed to be transparent between the crafts room and the corridor.

the serene and lush landscape below. The direct sunlight that reaches this space makes it attractive for relaxing or reading.

Atrium projects often contain wide balconies, which provide residents with the opportunity to sit outside their unit in a secure temperature-controlled environment, watching the activities below or visitors strolling on the opposite side of the atrium. In general, atriums provide a generous amount of natural light that makes other activities such as reading and needlework possible.

In some instances, the shape of a building facilitates observational possibilities. Single-loaded open balcony corridors wrap the inside edge of the U-shaped, three-story **Lille Gläsny** courtyard, in Odense, Denmark. Kitchen windows located near unit entry doors overlook the courtyard so that residents can watch visitors entering and exiting the building. Another technique used by **de Klinker**, in Amsterdam, and **Dronning Anne-Marie**, in Copenhagen, involves placing glass panels between lobby areas and a commercial kitchen. Residents can watch kitchen staff involved in a variety of meal preparation activities.

Finally, **de Drie Hoven**, in Amsterdam, is designed with the idea of linking together four L-shaped mid-rise buildings through a centrally located two-story atrium. This space contains most of the shared services and amenities and is considered the "town square" of the project. Because it connects all four buildings and channels pedestrian traffic, it is a vital, active, and interesting place to visit almost any time during the day or evening. A number of benches with built-in lamps and bookcases are located here in places that link circulation and activity spaces.

Spaces that Support Social Exchange

The most popular and heavily utilized social spaces in European projects were restaurants. They varied in scale but were designed as self-service cafeterias. Attendants assisted frail residents by carrying food to tables. Some restaurants were large multistory atrium spaces while others were broken into small, intimate rooms or divided into smaller areas by half-wall planters and screens. Projects with a bar, a billiards table, and a card playing area create informal places for residents and visitors to visit every day.

Projects that limited the number of residents taking a meal to between eight and ten use furniture, like dining room tables, to symbolize the family-like character of the setting. In the **Solgård** nursing home, in Tranemo, Swe-

FIGURE 4.29 The bar at the *Strandlund* project in Charlottenlund, Denmark, overlooks the entry to the main common area. Residents can take lunch here or visit any time during the day to relax, socialize, or read. A large fireplace out of camera range to the left, adds to the character of this space during the winter.

den, a fireplace and serving bar adjacent to each small kitchen table gives the space a cozy look resembling the scale and appearance of a residential kitchen. Policies that require staff to take meals with residents further solidify the family bond between staff and residents. The activity of taking a meal is one of the best opportunities for shared communication between staff and residents.

Although places for social exchange and passive observation are treated separately in this chapter, they are often used interchangeably. Spaces for observation can easily be transformed into places for informal socialization. Whyte (1980) introduces the notion of "triangulation" by describing how an activity shared by two people can become the catalyst for a conversation.

The most effective spaces that support both observational and social activities include entry areas inside and directly outside housing projects. Residents in these spaces are more at ease engaging in conversations that are easy to start because they can be terminated by the excuse to go out or back to a resident's apartment. In the outdoor space study referred to earlier, Regnier (1985), discovered the most popular sitting areas were adjacent to the entry of the building or along pathways leading from the entry to the neighborhood. In one project, an arcade with tables along the front of the building was parallel to a busy sidewalk. The arcade sitting area was separated by a screen of plants that provided shade and protection from the view of nearby pedestrians. Residents enjoyed watching the pedestrian activity without being seen. They selected a place to sit based on their interest in engaging in conversation. Those wanting more privacy moved to the ends of the arcade where they would not be disturbed. Those with more interest in casual verbal contact moved closest to the front entry.

FIGURE 4.30 The six different view streams that emanate from the decentralized dining room and snack area at the *Old People's Home and Health Center* at Oitti, Finland, make it a popular place to sit. Each view captures a different active scene. Some views are activated by staff, family, and residents while others depend on off-site movement or visitors. Together they combine to make this lounge the most popular place to sit, in the whole facility.

100 Percent Corner

The term 100 percent corner is often used to describe the sidewalk in an urban central business district that receives the highest amount of pedestrian traffic. In this context it is used to describe the portion of the housing project that attracts the most activity and use. Each project has its own 100 percent corner and occasionally two or three places in the project vie for the title. The major social challenge of designing group living arrangements is to create the circumstances necessary for places to become attractive in this way. One of the best examples of a heavily utilized 100 percent corner occurred in one of the lounges in the **Old People's Home and Health Center**, in Oitti, Finland. In this project, six factors combine to make this space popular. The following views are keyed to figure 4.30.

1. View Down the Corridor

The lounge space is located where the building's main corridor changes direction. Residents can see down the entire length of the corridor through a glass wall that separates the lounge from the corridor. This allows residents sitting in the lounge to watch others approaching.

2. Building Entry

Adjacent to the corridor and visible through the window wall to the left is an entrance to the building. Visitors, guests, family members, and residents entering and exiting the building are clearly visible.

3. Caregiver Office

Across from the entry is an office for nurses and the staff. A window wall separates this office from the corridor, making the activity inside this room visible from the corridor and lounge. Staff activities involve moving equipment, materials, and personnel in and out of this office during the day.

4. Food Preparation

Staff prepare meals and snacks in a small warm-up kitchen on one side of the lounge. Using the lounge for meals gives residents another purpose for visiting on a regular basis.

5. Adjacent to Skylighted Living Room

On the south edge of the lounge is a living room area located below a large pyramidal skylight. The living room provides additional daylight to the lounge.

6. Lounge Overlooks Walking Path

An exterior walkway runs under the lounge connecting the retail stores on one side of the project with a residential area on the other side. Windows facing the north overlook the path. People on foot and bicycle use this path on a steady basis during the day and provide residents with activity they can watch.

Activities, views, and overlooks like those at Oitti occasionally converge to create engaging places to sit. However, more often these settings are created by accident rather than by a concerted effort to select vantage points that are lively and interesting. Given their predictable nature, buildings should be designed to capitalize on influences such as those described above to make the place lively and engaging.

Summary

One of the main purposes for creating group living arrangements is the possibility they provide for residents to meet one another and create new friendships. Furthermore, some of the best urban spaces are those that allow individuals to passively engage the world around them by observing human activity and engaging in informal social exchange. Designers must recognize these two influences in establishing places for sitting and waiting. Utilizing preview techniques, capitalizing on the intrinsic vitality of the site and the context, and facilitating conversations through shared experiences are a few of the powerful techniques that can inspire architectural forms that are physically delightful and socially successful.

8. THERAPEUTIC DESIGN

Introduction

Assisted living should be viewed as a therapeutic environment where caregiving assistance and competence-building interventions encourage the highest level of independence. The physical setting can be an active contributor to this goal by supporting a therapeutic emphasis, or it can limit independence and stifle self-esteem. The word *therapy* in this context should be broadly defined but should include at its nexus concerns for physical exercise, mental stimulation, and opportunities for social exchange. When environments are designed to *only* overcome lost abilities, they are commonly viewed as "prosthetic" rather than therapeutic. Therapeutic environments challenge residents to exercise with the goal of renewing lost abilities or building new competency.

One major difference between a nursing home and assisted living is the attitude toward programs that merely support and those that challenge, thus keeping residents at a higher level of functioning and independence. Each resident's abilities are different and can change with age. The ultimate success of an assisted living program geared to keeping older frail residents out of an institution is management's commitment to a therapeutic philosophy.

The environment plays a key role by setting the stage for a therapeutic focus of mental stimulation, physical exercise, and social exchange. Ideas about how to make these experiences more convenient or central to the life of residents have important physical dimensions. Physical therapy can involve special equipment set aside in specific rooms or it can be designed into the building. Corridors, courtyards, and gardens can be designed to encourage physical therapy and exercise. Equipment can be located in a single obscure basement location, available only a few hours per week, or it can be placed in hallways and atriums accessible to residents whenever they want to use it.

Stimulating the mind through training or through a continuation of normal daily activities is an important dimension of cognitive therapy. Social interaction with caring friends allows exercise of another sort, which buoys the spirit, enhances mastery, and overcomes depression. Spaces designed to accommodate intergenerational activities and pet therapy tap differing dimensions of the human spirit, giving the setting an added measure of joy and affection.

Visible Physical Therapy

In many nursing homes, physical therapy is located in a windowless room and is viewed as a discretionary activity, requiring doctor's orders. The design of the building should make physical therapy and exercise a normal aspect of the lifestyle of the place. Increasing access to physical therapy equipment is an effective beginning point.

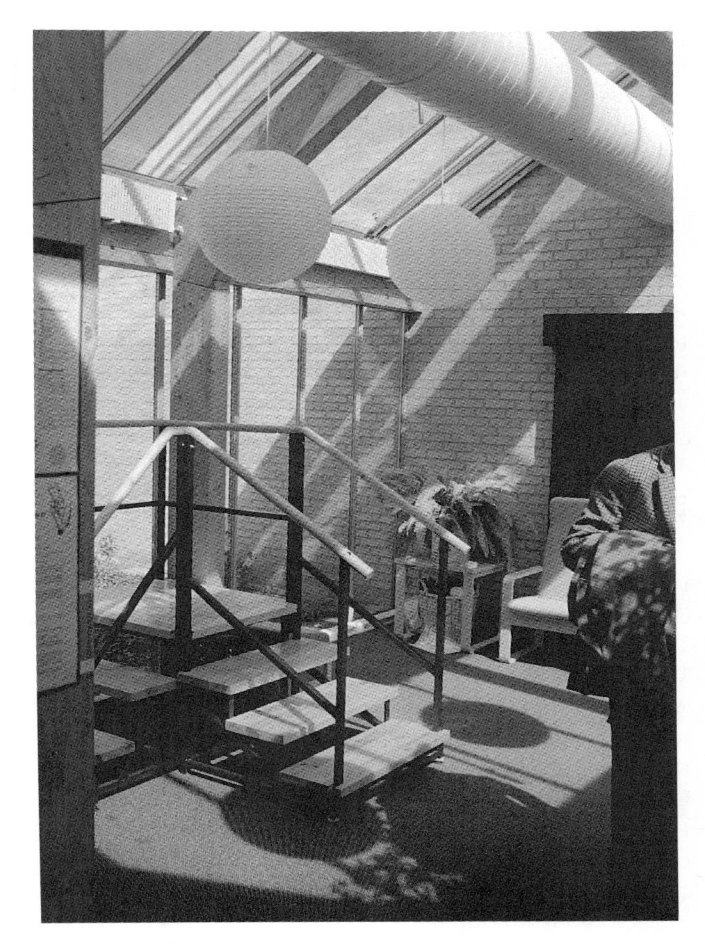

FIGURE 4.31 Decentralized opportunities for exercise can be found throughout Danish nursing homes and service houses. In the *Omsorgscentret Egegården,* in Soburg, Denmark, a clear glass greenhouse transition space between residential units and the community commons has a stair-climbing device residents can use by themselves or with staff assistance.

The most radical example encountered in northern Europe was the **Viherkoti** nursing home, in Espoo, Finland. Half of its large central atrium is devoted to a physical therapy and exercise space in full view of resident lounge and dining room spaces on upper floors. Portable equipment is located here on a resilient wood floor. The exercise area is separated from the rest of the atrium by portable partitions that can be moved when all of the atrium floor is needed for assembly purposes. The visibility of the space and its equipment is a powerful symbolic reminder to residents of the value of exercise and physical movement. Sophisticated equipment for specialized procedures and assessments is located in rooms adjacent to the atrium.

Another approach in the **Frederiksbroën** and **Egegården** projects, in Odense, Denmark, involves the placement of exercise bicycles, parallel bars, and stair climber devices in heavily used corridors. Their presence delivers an important symbolic message and makes it easier for residents to gain access to the equipment and use it at their own pace.

At **Mount San Antonio Gardens**, in Pomona, California, a former multipurpose room has been remodeled into a physical therapy and exercise space to facilitate access to therapy personnel and equipment. This remodeling of the building also has significant symbolic overtones, stressing to frail residents the value of maintaining upper and lower body muscle strength for ambulation and task manipulation.

Encouraging Walking

Physical therapists often use corridors to exercise residents. This is especially attractive during the winter when it is cold and often dangerous to walk outside. Walking for exercise is popular and beneficial. Atrium projects, especially those with gardens, allow residents to walk for exercise in an environment that resembles the outdoors. Projects with generous outdoor site areas like the **Vanhainkoti-Palvelukeskus Himmell**, in Pori, Finland, provide residents with places to walk for exercise and seating areas to rest and appreciate nature.

Designing buildings so that they encourage using stairs to go from one floor to another presents residents with a simple and natural way to exercise. Stairs also share a strong association with the residential imagery of the house. An elevator should be available for those with ambulatory impairments or balance control problems. The **Captain Eldridge Congregate House**, in Hyannis, **Sunrise** retirement community, in Fairfax, Virginia, and **Elder Homestead,** in Minnetonka, Minnesota, have open stairs in locations that encourage residents to use them. Elevators are generally placed to one side or behind the stairs in a less prominent location.

European designs differ considerably with regard to the design of exit stairs. In the United States, stair enclosures are covered with solid materials and are rarely used because they are located at the ends of corridors. In Europe, exit stairs are sheathed with transparent materials and often have delightful views to on- and off-site amenities.

Kids, Birds, Dogs, and Plants

Therapy can also involve accommodating a range of different types of stimulation.

Kids

Intergenerational programs involving young children are employed in projects in the United States and Europe. The **Fairfax Sunrise** retirement community is co-located with an elementary school and takes advantage of the presence of children through programs organized in the building's upper floor "tea room." Staff members remark that resident's bored by craft activities designed to stimulate small-muscle function become motivated and enthusiastic when they help children carry out the same tasks. The children become catalysts that transform the residents' role as a passive participant to an instructor. Helping children to master the task increases a resident's sense of self-worth.

In Lonköping, Sweden, city officials have pursued an aggressive plan of constructing family housing adjacent to elder care facilities. In these arrangements, day care for children is often located in the same building used by the elderly. In one project, windows along a busy corridor allow older residents to watch children playing in the day-care center. Informal observation seems more popular than structured interaction. Playgrounds visible from units are especially appreciated. In general, the potential of children to play an important therapeutic role has not been fully realized or explored. Environments that support observation and provide older residents with control over interaction are the most successful.

FIGURE 4.32 The most amazing example of visually prominent physical therapy equipment is the atrium of the *Viberkoti* nursing home, in Espoo, Finland. Half of the atrium space is devoted to physical therapy activities visible from resident rooms on the upper floors. This is not only convenient, but its prominence encourages residents and staff to use it more frequently. When the atrium is needed for other purposes, the temporary privacy walls located here can be moved.

Dogs and Birds

Programs that incorporate dogs, birds, cats, gerbils and other warmblooded animals report positive benefits that accrue from caged animals and "house pets" that roam corridors and common rooms. Outside an occasional barking dog, most criticism revolves around allergic reactions and problems with fleas. Assuming these problems can be managed, pets have the potential to animate a place and provide residents unconditional affection.

Swedish group homes effectively integrate animals into the home's daily routine. These dwellings for dementia victims closely resemble single-family housing in scale and character. Pets are attractive to dementia victims and are often successful in eliciting affective responses. There is much more we need to explore about how animals can be employed in therapeutic programs.

Most facilities are not designed well to accommodate pets. As a result, conflicts arise that could be mitigated by design adjustments. Many of the three-season porches used in United States assisted living facilities are uniquely positioned to deal with problems such as pet entry and exit, ventilation, and relative privacy for interaction.

Birds and Butterflies

Another potential of a site is its ability to attract birds, squirrels, butterflies, and other creatures to animate outdoor spaces. Few facilities seem to deal creatively with this possibility. Birdbaths, bird feeders, and birdhouses placed as decorative elements in courtyards have been the most common ideas. Although facilities need not be transformed into animal sanctuaries or zoos, creative thinking about how to include animals for therapeutic purposes is warranted on the basis of initial encouraging reactions.

Plant Materials

Greenhouses attached to interior corridors or constructed as freestanding buildings are popular in a number of assisted living buildings. In frost belt states where natural landscapes are unavailable for as much as half the year, they play an important role. Solariums and sun rooms are a tradition in institutions where older residents remain bedfast and unable to go outside. Bringing the natural landscape inside and making it accessible is the concept behind the European garden atrium.

European facilities in general seem more sensitive to the therapeutic potential of plants. Resident rooms in Danish nursing homes frequently open onto an outdoor patio and ideas like raised planter beds are often employed to make the color, texture, and aroma of plants easier to experience for the wheelchair-bound. Swedish group homes often contain herb and vegetable gardens to encourage a range of normal daily activities.

However, many facilities lack an understanding of the meaning of plants and their therapeutic potential. The possibility of using plants to stimulate the senses, trigger past associations, and generally soften the appearance of the environment is greatly underexplored.

Occupational and Physical Therapy

The most impressive therapeutic facilities are the ergotherapy (occupational therapy) and physical therapy rooms of the Danish plegehem. Their commitment to engaging older residents in activities that stimulate a positive mental attitude and a therapeutic physical response is truly astounding. The United States and most other cultures conceptualize occupational and physical therapy as part of a medical model rather than a lifestyle component of housing. The result is a weak commitment to this activity. Older residents at this point in their physical and mental decline *need* focused and concentrated effort to maintain or build abilities. When health and wellness are recognized, it is seen primarily as a "programmatic component" of housing rather than creatively explored in the architecture of the building and the surrounding landscape. Settings should be designed to stimulate exercise and encourage a range of activities.

Another significant dimension of the Danish system is its desire to make occupational therapy a "normal" activity, similar in nature to the routinized responsibilities of working people or the housekeeping responsibilities of homemakers. Ergotherapy, like work, is a productive enterprise that gives residents a sense of satisfaction. The activities pursued have significant therapeutic benefits and result in the creation of a product that has meaning and value. Weaving is the most common European activity that reaps this combination of benefits. Ergotherapy creatively pursues the notion that older residents have a responsibility to stay active and engaged.

FIGURE 4.33 The courtyard of the *Rygardcentret* in Gentofte, Denmark (Hoglund, 1985), is used for walking therapy. A collection of different floor materials (brick, concrete, turf, wood) installed here challenges different leg and foot muscles.

Swimming Pools and Saunas

Swimming and water-related exercise therapies have long been recognized as particularly well suited to the needs of older people who benefit from the warmth of the water and the nonjarring, smooth, and natural resistance to movement that it provides (Cooper, 1976; Racavas, Morrison, and Hurley, 1977). However, two issues impede swimming therapy. The first is a cultural concern. Many older people have not learned to swim and do not feel comfortable in a swimming pool. Older people may also be overly concerned about the body image effects of aging and may feel ill at ease in a public setting. The second issue concerns cost. Swimming pools involve expensive first-time capital investments and incur substantial continuing operating and maintenance expenses. These two impediments have strongly impacted the popularity and use of swimming pools in the United States. Smaller projects cannot afford these initial fixed or ongoing costs.

Eight projects with swimming pools were site-visited in Denmark and Finland. In most cases an agreement with local government had aided the financing of the swimming pool, opening it to other groups from the surrounding neighborhood. These groups included children and young adults with physical disabilities, as well as other older people beside the residents. Interestingly, two of the projects were equity-based cooperatives that entered into a public-private partnership to finance and manage a jointly owned swimming pool. Swimming pools in Finland are always linked with saunas. In the Finnish culture, older people often use the sauna on a weekly basis as a substitute for bathing. Thus, a number of community residents are motivated to use these facilities for personal hygiene purposes. Well designed for wheelchair access, each pool has a sunken pit on one side to facilitate the involvement of an instructor or therapist.

Swimming pools installed in basement locations are often cleverly daylighted. Older people are more likely to suffer hearing losses and have difficulty hearing oral instructions from a therapist. Wood or concrete sound absorption panels in the ceiling were used because of the proliferation of hard surfaces and the humidity that together render traditional sound insulation useless.

Mixed-Use Buildings

The architectural program of the building can include spaces and equipment that enhance resident well-being as well as the therapeutic potential of the setting. Augmenting the building program to include access to health services can enhance the project's therapeutic potential, but often at the expense of residential scale.

However, some additional changes can be minimal. At the **Degneparken** project in Dianalund, Denmark, six long-term care beds were added to the first floor. In this rural community, the need to provide respite care and more intensive health care for individuals moving from the hospital back into the community required a small infirmary.

The **Kotikallio** service house, in Helsinki, was designed to serve 87 adjacent condominiums and 140 older people living in four buildings in the surrounding neighborhood. The four-story building includes a thirty-five bed rehabilitation hospital on the top floor. This program addition increased the amount of equipment and personnel available in the physical therapy area. The outcome is a service center with sophisticated physical therapy services that can be used by residents in attached housing.

The **Old People's Home and Health Center**, at Oitti, Finland, originated from a competition that combined an outpatient care center with two levels of supportive housing (*The Finnish Architectural Review*, 1987). Adding housing to the program increased the flexibility of the setting, making it easier to treat medical problems on-site and provide more extensive therapies for residents.

The **Runby** service house, in Upplands Väsby, Sweden, is developed above a restaurant, library, doctors' offices, and clinical treatment spaces. When medical offices are located below housing but with separate street access and identity, they need not affect the character of the housing.

Summary

The objectives of assisted living differ from those of nursing home environments. Programs in assisted living are dedicated to building resident competence and restoring lost abilities. The design of the building should support a therapeutic lifestyle that stresses physical exercise and activity, cognitive development, and social interaction. Design can aid therapy through the convenient location of special therapeutic spaces and by the way they are given visibility within the building. The environment should assist programs that involve intergenerational exchange, pet therapy, and physical therapy.

FIGURE 4.34 The atrium at *Kuuselan Palvelukoti,* in Tampere, Finland, is designed to accommodate a swimming pool that overlooks the atrium's lush landscape. The light-level difference between the naturally sunlit atrium and the darker enclosed swimming pool creates a one-way mirror effect that preserves the privacy of swimmers from outside viewers, but maintains clear visibility of the garden from the inside.

9. LINKING AND CONNECTING SPACES

Introduction

Most United States nursing homes look like a series of long, extruded double-loaded corridors linking small boxlike rooms. Eight-foot-wide corridors are specified so two beds can pass one another, connected to rooms through solid fire-rated walls. Typically, each room accommodates two unrelated individuals and rarely do patient rooms have doors that open to the outside. The minimum connections between spaces and the lack of spatial variety create monotony and a depressing sameness. The rigid geometry of corridors designed to efficiently connect the nurses' station to each patient room produces symmetrical configurations that resemble a prison more than the hospital form from which it was derived. These building forms lack spatial hierarchy, minimize previewing opportunities, expose the private world to the semipublic corridor, create difficulty in orientation and wayfinding for patient and visitor, and add to the rat-maze-like complexity of the setting.

Nursing homes lack almost all the normal references one might encounter in a residential environment. A lack of creative spatial connections occurs between the indoors and the outdoors, between corridors and common activity rooms, and between residential corridors and patient rooms. When necessary connections like doors occur, they operate for the convenience of staff, rather than the privacy needs of the older person. For example, doors are left open most of the time to monitor resident behavior. This doesn't allow the resident control over privacy and ironically limits rather than enhances verbal communication with staff. Access to the outdoors from the unit is often considered an unnecessary hazard because of wandering behavior. However, sill heights are often high and landscape planting nonexistent, making the view from the resident's bed minimal.

Dwelling units in nursing homes are isolated from the world, and alienated from what we consider normal and necessary in residential housing. Social and recreational spaces are often equally disadvantaged. Required fire code separations between corridors, common rooms, and separate floors, discourage the use of atriums, open walls, pierced planes, deep portal entries, half walls, columns, planters, and bookcases as spatial connectors. Fire codes often require that windows from corridors into these common meeting rooms be minimized in size or outfitted with rated wire glass, further denigrating the character of these spaces.

The result is a collection of anonymous rooms disconnected from one another. Opportunities to see into spaces, "previewing" them before entering, are minimized and the ability to see through spaces to comprehend a pattern of linkages for wayfinding purposes is extinguished. When spaces link, overlap, and connect, numerous benefits affecting use, habitability, and sociability accrue. When they do not, these potential benefits are lost.

Spatial Linkages

There are many ways to spatially and visually connect spaces that use doors, wall openings, and windows as basic building blocks. The most intriguing are materials that separate rooms by partially obscuring the visual connection between spaces. Plant materials are one popular way to achieve this. Restaurants often use planter box dividers, thirty-six to forty-two inches in height, adjacent to tables and chairs to achieve the feeling of privacy while allowing the room to feel relatively open. In **Vickelbygården**, in Skärblacka, Sweden, a low planter forms the foundation for a wood mullion grid that stretches from the floor to the ceiling, separating the corridor from a dining room and lounge. Viny plant materials espaliered on the wood grid further obscure the view and introduce a soft green texture, which varies in visual complexity and density. At **Palvelutalo Esikko**, in Turku, Finland, a similar concept has been employed to separate an open balcony corridor from an adjacent atrium. Wire stretched in a vertical pattern over the opening allows plant vines to create an ephemeral and partially transparent edge. An additional benefit of this application is its view of the soft four-story, ivy-covered "green wall" from the first floor of the atrium.

Obscure glass can also be used to link spaces. At the **Old People's Home and Health Center**, in Oitti, Finland, walls of sandblasted glass and glass block are used interchangeably with clear glass and cased openings to produce translucent and transparent spatial linkages that accommodate privacy when needed.

Another popular approach has been to use double-hung exterior windows to link together interior spaces. The **Captain Eldridge Congregate House**, in Hyannis, introduced these between-dwelling unit kitchens and an adjacent atrium. At the **Annie Maxim House**, in Rochester, Massachusetts, they are used between corridors and the kitchen, laundry, and living room. The advantage of these windows is that they can be opened for ventilation and communication or closed to isolate sound and secure privacy. At the **John Bertram House**, in Salem, Massachusetts, an informal country kitchen dining room was created within an old bay window alcove. A new enclosed corridor addition to the exterior of the building transformed the bay window into a window wall overlooking the corridor. Three large double-hung windows give those at the informal kitchen table placed here excellent visibility as well as a moderate degree of privacy, creating the necessary physical conditions for a highly successful social space.

FIGURE 4.35 Plant materials trained along horizontal bars partially obscure the edge of the residential corridor at the *Palvelutalo Esikko,* in Turku, Finland. This is a very effective in creating a connected but separate feeling between the atrium and the corridor.

Half Walls and Deep Portal Entries

Another effective way to create edges to spaces without making them feel closed in is to use half walls, which define space by demarcating edges. Half walls can also be thickened to include bookcases, storage cabinets, or built-in planter boxes. When this wall becomes larger than twelve to fifteen inches, another curious perceptual phenomenon occurs. The interstitial space between the two adjacent rooms takes on definition. This is most apparent in doorways where deep portal entries allow someone entering the room to preview the space from a "mini-vestibule" without making a commitment to enter the space. These types of spaces are common in eighteenth-century housing (Moore, Allen, and Lyndon, 1974). Such transitions can be enhanced by lowering the ceiling height or changing the floor or ceiling materials.

At **Woodside Place**, in Oakmont, Pennsylvania, a sitting area in the middle of the building adjacent to a fireplace is separated from the surrounding corridor by a twenty-four-inch-deep built-in bookcase and storage area, which serves as a half wall. This built-in casework creates an intimate space for conversations, while allowing daylight to penetrate, and permitting views to the grounds.

Another technique used to define space, in the **Elder Homestead** project, in Minnetonka, Minnesota, involves freestanding columns used in conjunction with casework. These have been used as portal elements in the middle of a long corridor to perceptually foreshorten its length and as an entry element between the entry and the first-floor living room. Casework platforms for these columns allow them to effectively define space.

FIGURE 4.36 A half wall using a wooden grid design at the *Vickelbygården* nursing home, in Skärblacka, Sweden, creates a sense of separation and connection between the corridor and a dining room. Note the fireplace at the corner of the room. American fire codes would forbid the use of an open fireplace in a nursing home lounge.

Conclusion

Connecting and linking spaces can transform a boring, continuous, and disorienting collection of self-contained rooms and corridors into a varied composition of places that overlap and relate to one another. A variety of devices such as half walls, planters, portal entries, screen walls, bookcases, partial height partitions, atriums, and openable windows can be used to achieve a range of desired effects. Linking and connecting spaces facilitates previewing, provides a better sense of orientation, adds to spatial variety, and allows borrowed natural light to reach farther into the building. When these connections work well, they provide residents with more control over the encounters they create in shared spaces, thus increasing the potential for social exchange.

10. INVITING THE FAMILY

Introduction

Moving from an independent dwelling unit into a group living arrangement can be a traumatic experience for many older people. To some, it symbolizes a move toward institutionalization that implies a reduction in individual freedom, and the imposition of rigid, impersonal rules. To others, it connotes a change from a highly controllable private life in a single-family house to a lifestyle that involves collective social participation. To families, the move may shift control of the older person's welfare away from family decision making. Day-to-day responsibilities for care management move from the jurisdiction of the family to the domain of the formal provider, who by law must take responsibility for the older person.

Most facilities do not creatively involve families. For example, families are rarely active participants in comanaging social, health, and nutritional services. Families accept institutionalization as a process that narrows their responsibility and their overall role in caregiving. Facilities often reinforce this by treating family participation as a useful but discretionary activity. Instead of being viewed as a significant source of affective support and inspiration to the resident, it is often tolerated but not encouraged.

Some families are not interested in participating while other residents are without family members and therefore have few family care options. The rigid, segregated nature of the system, however, discourages many family members who are motivated to continue an active and normal relationship with their parent. Building form and management policies are often major impediments. Narrowly conceived regulations limit formal-informal caregiving partnerships, creating major legal impairments in some states. However, the worst problems are caused by conventional thinking and the continuation of bad habits.

Facilities are also not designed around places, spaces, and activities that attract or invite the family. One major impediment is the double-occupied room, common in most nursing facilities. Single-occupied rooms with full bathroom accommodations and a food storage or preparation space promote privacy and encourage family visits and overnight stays.

Facilities must involve families in more creative ways than they have in the past. One way to rethink this situation is by examining other models of social care, such as day care for children. In these settings, family participation has been an important and integral aspect of the way these settings have been regulated and operated. Rules abound but the ability to accommodate family needs and desires is not compromised. Home-based care is another integrated family model, which coordinates formal and informal resources to keep the older person functioning independently outside an institution. The radical idea of giving the family the shared responsibility to provide services, within a setting where formal help can be efficiently ordered when necessary, expands the parameters of thinking about how to design family-centered models of assisted living.

Role of the Family in European Housing

European systems seem less able to provide creative insights with regard to family care, partly as a result of their success in providing adequate financial support through nationalized health insurance. Northern Europeans pay the highest taxes in the world for a system of excellent cradle to grave care. This has ironically preempted efforts at developing creative family care options. The family interaction that does occur is largely social in nature. Family members visit, transport residents to the doctor, and occasionally help with major cleaning tasks; but they offer little instrumental care because the system provides it. The system has established high expectations for services, operating without the need for significant family input.

However, European models are increasingly focused on self-maintenance models that coax older people to exert maximum participation in activities of daily living. Care attendants involve residents in self-care tasks, such as bathing, by having them do as much for themselves as they are physically capable of doing. Such a model of care provision is more difficult to implement than doing the work for the resident, but the benefits of this approach are great. Residents feel more confident, with a greater sense of self-esteem through practicing behaviors that increase their overall abilities. This is a therapeutic approach to providing care that builds competence rather than taking it away by doing too much for residents.

Home Care Assessment Models in the United States

In the United States, the increasing cost of care is moving consumers toward systems that encourage greater participation by family members. At the **Rackleff House**, in Canby, Oregon, residents are assessed using a community care model. Significant resources from residents themselves, their families, and other residents are considered as sources of support. Facilities for laundry are made available to family members, meals are offered to them at discount prices, and they are encouraged to help in any way they can. Care plans are devised by consulting the older person and the person's family in specifying needed services.

Controversial models like **Rosewood Estate**, in Roseville, Minnesota, reviewed in detail in Chapter Five, are based on a home health care philosophy that has been transposed into a group living environment. In some states, this approach violates state statutes that consider providing this form of care as operating a facility without a license. In this model, the potential level of health and personal care available to each resident is unlimited. Only the ability to pay limits what can be accomplished. This type of caregiving system is a true partnership. Care plans allow family members to share the burden of care. Families can do laundry, organize and administer medications, supervise baths, and help their family member to the dining room. Care needs are assessed and monitored by a case manager, and residents pay only for the care they need. Case managers ensure that the appropriate level of care is covered by either a family member or a formal care attendant.

Partnerships distribute work between formal and informal sources. For example, families can set up medicines in a weekly pillbox while staff members take responsibility for reminders. A daughter can help her mother take a bath several evenings each week while staff retain responsibility for toileting assistance during the night. Intervals of care, sold in fifteen-minute segments, make it easy to choose what help will be provided formally and what will be provided by the family. The model challenges residents to perform to a maximum level, invites the family to participate in a broadbased and flexible way, encourages helping behaviors between residents, and provides care in the privacy of each resident's unit. Private self-contained dwelling units also facilitate hospice care, which intensively involves family members in a joint effort with health care professionals.

The family can also exert influence and take responsibility more easily in the Swedish group home model. Its small size allows families to host events and take a very active role in setting policies and scheduling activities. In group homes, care attendants are trained to work with family members. They council and advise on effective

FIGURE 4.37 Swedish group homes for mentally frail older residents use a range of activity of daily living therapies to stimulate residents. Laundry rooms are centrally located and residents participate by collecting laundry, placing it in the washer and dryer, folding it after it has been cleaned and putting it away.

FIGURE 4.38 The fireplace at *Woodside Place,* in Oakmont, Pennsylvania, establishes a homey character for the living room that residents and their family members enjoy. The wide half-wall counters that separate the living room from the adjacent wandering pathway, give it visibility but also ensure privacy.

techniques for interacting with demented relatives. Resentment, misunderstanding, and a general inability to accept a dementing disease can keep family members emotionally distant.

Role of the Physical Environment in Family Involvement

One factor that affects family involvement is the privacy of the dwelling unit. Family visits often occur in the dwelling unit. When the room is shared with another person, it is difficult to carry out a private conversation, thereby making visiting unsatisfying. The Danish dwelling unit in a typical service house is a one-bedroom unit of approximately 575–600 square feet. A full bathroom and a complete kitchen facilitate independence and encourages family visitation. Services are provided on an as-needed basis, including nursing care.

Providing places within the building for family members and residents to meet and socialize is another way of encouraging visiting behaviors. European service houses provide many choices, because they are physically connected to a range of community spaces like restaurants and cafés. In American assisted living environments, the enclosed porch has become an attractive room for family meetings.

In the United States, **Peachwood Inn**, in Rochester Hills, Michigan, has focused on creating small group social spaces in alcoves adjacent to major pedestrian thoroughfares. They have also opened the restaurant to outside customers in an effort to encourage family members and visitors to think of the facility as a potential alternative place where everyone can eat.

Finally, many facilities seek opportunities to host the families of residents for picnics and barbecues. Memorial Day, July 4, and Labor Day are important summer holidays historically associated with family picnics and events in the United States. Thanksgiving and Christmas are also traditional holidays that have family involvement. Both outdoor and indoor spaces should be designed to accommodate these family events. In the **Viherkoti** project, in Espoo, Finland, an atrium is contiguous with an open crafts and reading room. These rooms in turn open onto an exterior garden. Special events involving families allow all three spaces to be connected, thus facilitating larger groups.

Another way of helping families take care of older relatives is to make services like respite and day care available to older frail people living with family caregivers in the community. European housing models almost universally adopt this community-oriented approach. They view their role as providing a range of supportive services for residents and older people in the community.

Fellow Residents as Family

Relatives constitute the most conventional conceptualization of family. However, as residents adapt to the social environment of a group housing setting, they often create informal helping relationships that constitute another type of family network. Smaller facilities that encourage interdependent helping behaviors are the most likely to experience this bond of affiliation. At the **Captain Eldridge Congregate House**, in Hyannis, management believes "family" togetherness is a powerful motivating factor that keeps many residents independent. For example, a blind but ambulatory person will help a sighted but disabled resident and vice versa. Networks of interdependence occur when residents who know one another recognize opportunities for reciprocity. In general, these exchanges extend from friendships. They occur somewhat randomly, aided by propinquity and shared interests. Interdependence also occurs with married couples, whose problems are offset by the compensating competencies of partners.

FIGURE 4.39 Small food preparation kitchens are common in the units of dementia residents. Ranges are often designed to be switched from a remote location so that family members or residents can use them with supervision. Swedish building standards require that they be included in each resident's unit.

Communal and Co-Housing Projects

Some projects have gone to great efforts to establish a shared basis for collective living. The Danes are well known for their co-housing projects (Franck and Ahrentzen, 1990). Although relatively common, few are designed to care for the needs of older frail people. One that does is the **Skelager** project, in Åarhus, Denmark. In this project, six L-shaped one- and two-story clusters are linked together in courtyard forms. Eight dwelling units are clustered around a shared living room, dining room, and kitchen. A part-time facilitator/caregiver takes managerial and organizational responsibility for shopping, and food preparation. Services, assigned as required, are provided by home care workers. What makes this project unique is that residents make a work contribution to the household, consistent with their abilities. At the core of this program is a strong self-management thrust that inventories the strengths and service need of the eight residents living together in each cluster. Assessments form the basis for a task management plan that is updated periodically as residents' abilities change and as new replacements join the group.

In the Netherlands, there has been a great deal of experimentation with forms of communal housing for the younger old (Toneman, 1990). Projects vary in size, but most have fewer than twenty-five units. The **de Boogerd** project, in Hoorn, is a relatively large project of eighty-seven units, which establishes work tasks around informal agreements between residents, negotiated by an elected nine-member resident committee. Ten task-oriented working groups take responsibility for gardening, cleaning public areas, social activities, and other management concerns. Money saved as a result of resident-contributed labor is returned in the form of lower monthly rental fees. Although each unit has a kitchen, small groups of residents use a common kitchen and dining room on a daily informal basis, where they share food purchase and meal preparation responsibilities. The Dutch communal projects generally do not have any formal mechanism for dealing with frail residents other than through reliance on home care services. However, the communal nature of the setting encourages residents to help one another and to augment formal support with informal aid and assistance.

Friendly Neighbors Through Community Service

Although we often view older frail people as consumers of service, they can also be significant producers of service. The intergenerational program described earlier, in which older residents and children benefit from one another, is a good example. Numerous housing projects work with community organizations contributing time and effort. Some of the most effective occur in small-town and rural settings, where the townspeople and elderly residents know one another. In Nr. Nebel, Denmark, the residents of **Baunbo** sew uniforms for the local high school hockey team. During high school graduation, they offered the enclosed street that links dwelling units to the local high school for a graduation party.

Selling products from ergotherapy, such as rugs and jewelry, is another way of relating to the community. Most service houses do this to cover the cost of materials, charging a modest profit to compensate the older people who produced the items for sale. The **Fabriken** project, in Linköping, Sweden, has a sheltered workshop which produces toys and crafts. Adjacent to the workshop, facing a busy street, is a retail store that sells items made in the studio.

Summary

Inviting family members can involve creating systems for sharing caregiving responsibilities and designing social spaces and outdoor areas where family members and older residents can spend time together. Private dwelling units that accommodate the family are also important. Group living arrangements are constrained by laws and regulations that limit the sharing of caregiving responsibilities with family members. Because of this, many older people fear a move to group housing may further erode family ties.

Creating a feeling of family among residents is another way to interpret its meaning within this context. Communal housing does this well by engaging older residents in jointly managing the household. These responsibilities can also lead to friendships that can allow residents to help one another. Inviting family members and creating networks of interdependence are primarily controlled by management policy, but the design of the building can also play an instrumental role in the success of these goals.

11. REALM BETWEEN INDOOR AND OUTDOOR SPACE

Introduction

Spaces that occur between the protection of the conditioned interior of a building and the natural forces of the exterior environment have special significance for older frail people, because they allow the protection of the dwelling to extend into the surrounding landscape. This increases resident opportunities to view, experience and appreciate the outdoors. In thinking about this realm, two spatial typologies emerge.

The first includes interior spaces at the edge of a room, thrust into the landscape and exposed to the sun or an interesting view. Spaces like greenhouse windows, solariums, enclosed balconies, three-season porches, bay windows, and the Dutch erker fall into this spatial category. These interior spaces exist within the secure and protected domain of the building, allowing residents to directly engage in observing on- and off-site activities.

The second typology includes exterior spaces protected by a dwelling roof or extension that covers and defines the space. These spaces are closely "coupled" with the exterior edge of the building, affording easy access to the inside and providing the older person with a sense of security and protection. The most common example of this spatial type is the front porch, although arcades, trellises, covered patios, and balconies fit within this category.

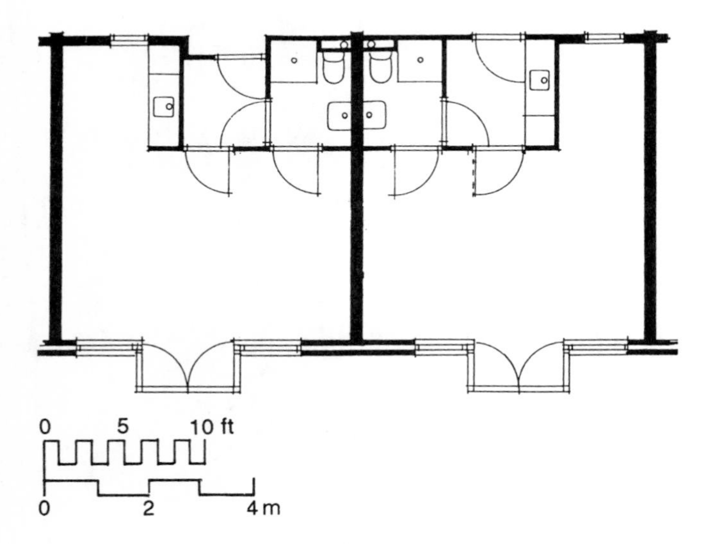

FIGURE 4.40 The square bay window the Dutch call *erkers* allow residents excellent view access to street activity below. Most residents treat the interior alcove as a sitting area where they place an easy chair. The side windows of the erker have low sills to facilitate views.

Drawing a line between these two types of space is somewhat difficult. When a screened porch is enclosed and protected from the elements, it takes on the characteristics of an interior space. Perhaps the line between these spatial typologies could be drawn when a porch enclosure is conditioned and protected from the elements. Another definitional aspect of these spaces is their need to be contiguous with the building. Once a space breaks away from the dwelling, it becomes a separate pavilion. For example, a greenhouse window could qualify as an interior edge space, and an attached lath house might be considered an exterior edge space, but when a greenhouse or gazebo is located in the middle of the garden away from the building, it no longer possesses the edge qualities that place it in the realm between indoor and outdoor.

Another aspect of these spaces is their placement within a public to private continuum. Does the space have private, semiprivate, semipublic, or public access? Attached to the building entry, the space is considered public. As part of a common recreational or social space it becomes semipublic. Located on an upper floor, accessible to a single cluster of units, it is considered semiprivate. Finally, attached to the unit in the form of a balcony or patio, it is considered private. Although its place in this continuum does not affect the basic concept of the indoor-outdoor space, it does affect its form, size, and use pattern.

Older people with disabilities or memory disorders often find it difficult to walk around the site or neighborhood. For them, these spaces provide a window to the world that allows easier access to outdoor experiences. They support solitary activities like reading or social activities like card-playing equally well. Porch and arcade spaces are often designed to function as circulation areas as well as places to sit. When so configured, they allow residents to initiate conversations that can eventually lead to friendships. These exterior attachments can also increase the variety, complexity, and interest of the building.

Interior Edge Spaces

One of the most interesting European examples of an interior edge space is the Dutch erker. An alcove attached to a dwelling unit living room, the erker juts eighteen to twenty-four inches over an adjacent street like a square bay window. Most have a width of four to six feet. Erkers are popular on the upper floors of urban housing, often designed with glass on three sides so residents can see up, down, and across the street. They are particularly popular when located next to heavily traveled streets and squares that afford enhanced opportunities for viewing activity. An additional side benefit is the visual pattern they establish when used as part of a continuous facade. The **Flesseman**, **de Klinker**, and **de Gooyer** projects, in Amsterdam, employ erkers to extend units.

Eaton Terrace II, in Lakewood, Colorado, also includes a small alcove adjacent to the edge of the living room. This area is tiled, allowing it to be used as a greenhouse for plants, a place for caged birds, or an extension of the living room. The tile floor covering and the configuration of this space jutting out of the building envelope give it a separate feeling. Low sill heights on the two sides invite additional light into the unit and facilitate views to the ground plane from this mid-rise project.

Functioning greenhouse enclosures located on the sunny side of a project can foster gardening activities and daylight interior spaces. The **Elder Homestead** project, in Minnetonka, Minnesota, uses a two-story glass wall to bring sunlight into an open greenhouse area located adjacent to a main circulation corridor. The greenhouse adds variety to the corridor and provides residents with an attractive experience on the way to the dining room. In the **Vanhainkoti-Parvakeskus** project, in Pori, Finland, the top floor has been used as a solarium greenhouse. Residents and their families socialize here against a lush backdrop of tropical plants.

The **Fyensgadecentret** project, in Ålborg, Denmark, has sheathed the corner of its dining room in glass. The location, adjacent to the entry of the site, provides residents with a cozy place to sit at a table and watch the surrounding street traffic. At the **Frederiksbroën** project, in Odense, Denmark, the service house restaurant has a greenhouse alcove that accommodates fifteen to twenty people. This portion of the restaurant can be closed off to form a semiprivate meeting place for clubs. The view is of an adjacent landscaped courtyard.

FIGURE 4.41 The erker windows at the *Flesseman Center*, in Amsterdam, facilitate the observation of activity down the street to Nieuwmarkt Square. The erker is a common unit feature in urban housing for the elderly. Note the French balcony doors that open inward and the retractable awning which can be used to control sunlight.

Several projects utilize glass enclosures, transparent attachments, and skylights to mark the front entry or to connect the project with the street or the surrounding landscape. In the **Gulkrögcentret** project, in Vieje, Denmark, a two-story glass entry connects the residential portion of the building with community spaces. A half dozen trees located within this entry space give it a garden character, encouraging its use as an informal social space for taking coffee or waiting for a ride. At the **Egegården** project, in Soborg, Denmark, a transparent glass transition space between the community service building and the residential wings daylights this area. In addition to introducing natural light, the views from this

FIGURE 4.42 An interior greenhouse space at the *Omorgscentret Egegården,* in Soburg, Denmark, is large enough to accommodate tables with bookcases. Residents can watch on-site activity from this light-filled space.

space provide useful cues for orientation. The space also houses library books, table games, and exercise equipment.

In the **Kvarteret Karl XI** project, in Halmstad, Sweden, an eight-foot-wide glass-enclosed, single-loaded corridor has been transformed by residents into a semiprivate enclosed balcony. Residents have moved furniture, plants, books, and chairs into this space, and use it as an extension of their dwelling unit. Each resident has treated the space somewhat differently, adding variety and individuality to the experience of walking down the corridor. This five-story portion of the building overlooks the adjacent central shopping district. It appears transparent at night, revealing to the neighborhood a varied mix of sitting areas, each decorated in a different personal way.

Another unusual interior edge space is the *serre*, a French term used to describe a balcony sun room sheathed in glass on all four sides. The **Bouwacher** and **Kruistraat** projects, in Utrecht, the Netherlands, feature this type of space. Project residents use the serre as either an interior or exterior room. They either carpet the space, treating it like an open and transparent sun room, or use it like a balcony with exterior lawn furniture and plants. The transparent nature of the room allows sunlight to penetrate into the living room behind.

Exterior Edge Spaces

The most common example of this type of space is a porch. Many United States projects utilize porches because they look friendly and provide an opportunity to connect the housing to the street in a direct way. The **Captain Eldridge Congregate House**, in Hyannis, **Elder Homestead**, in Minnetonka, Minnesota, and the **Sunrise** retirement community in Fairfax Virginia each use the porch to establish a sense of scale while providing residents with a place to gather and watch off-site activities. The **Fredericksbroën** project, in Odense, Denmark, has used an arcade as a buffer between the restaurant and a garden located at the rear of the site. Tables and chairs located here give residents a choice of taking a meal or snack outside.

Exterior single-loaded balcony corridors can also be an exterior edge space. At the **Sportsveien** project, in Oslo, small sitting areas situated on one side of the corridor are located adjacent to unit entries. At the **Moerwijk** project, in The Hague, three conditions of enclosure have been pursued with regard to the design of a continuous single-loaded corridor. At one end the corridor is left open above a solid lower rail. In another area it is fully enclosed in glass. A third portion uses glass as a windbreak with a twelve- to eighteen-inch open air gap at the top that allows breezes to enter the corridor. The perceptual differences among these three conditions are

FIGURE 4.43 Attached back porches at the *Raufosstun Eldresenter,* in Raufoss, Norway, surround a stand of birch trees in a courtyard space reminiscent of the old eighteenth-century tuns or courtyards.

pronounced. The partially protected option seems most desirable because it protects residents from harsh elements but retains the feeling of a lively open corridor.

A further interesting use of the covered porch idea is at the rear of the unit. At **Raufosstun**, in Raufoss, Norway, small covered patios protect residents and provide a place to sit around a fifty-foot-square landscaped courtyard. This treatment at the edge of each unit gives the courtyard a friendly and comfortable perimeter and the impression that each unit is entered through the back porch rather than from an enclosed corridor.

Summary

Two spatial realms exist between inside and outside spaces. Those located inside the dwelling that overlook the site include greenhouse enclosures, bay windows, window seats, and enclosed porches. Outdoor spaces located adjacent to the edge of the building but sheltered by the building roof include arcades, porches, and covered patios. Both of these spaces are attractive to older frail people with limited mobility and energy because they allow older residents to appreciate and experience the surrounding landscape from a protected domain. These two spatial types provide a multitude of positive benefits that include increasing the variety and massing possibilities of the building, reducing the scale of the project from the street, stimulating social interaction, and facilitating the observation of surrounding activity.

12. DWELLING UNIT FEATURES

Introduction

The dwelling unit is the most important setting for many older frail people, because they spend the greatest amount of time here. When comparing assisted living with nursing homes, some of the greatest symbolic and functional differences emerge at the dwelling unit level. The assisted living dwelling unit should be large enough to accommodate a full bathroom, a small kitchenette for food storage and preparation, and enough space for a family member to stay overnight. Without a kitchenette and bathroom, the dwelling unit cannot be considered complete. It becomes akin to a hotel room. The symbolism attached to a "normal" dwelling unit is very important. Each dwelling unit should also be private, with the capability of being locked. Rooms without doors that lock are controlled by the staff, not the resident. Such lack of privacy is an attribute of institutional settings.

Adaptability is also an important design criteria. Residents with differing problems and needs should be able to stay in their unit as they age. Dwelling units and helping services must adjust to personal needs as they change. English architect Martin Valins refers to the current lack of adaptability in dwelling units by stating that most older people live in "Peter Pan" housing, dwelling unit designs assume occupants will never age and their needs will never change beyond what was required as a younger able-bodied adult.

Designing the dwelling unit to meet wheelchair-based handicapped standards is not enough. Many older frail people seek to maintain their ambulatory abilities as long as possible. Rather than relying on a wheelchair, they use walkers, canes or assistive devices, or use furniture or bathroom fixtures to steady themselves as they walk. Making a room large enough to accommodate wheelchairs can make it less intimate and can complicate the trip from the bed to the bathroom in the middle of the night. The environment should be designed to accommodate a variety of potential use patterns, including wheelchair use. A larger wheelchair-adaptable environment can be filled with storage units and fixtures that make it more intimate, convenient, and accessible for frail ambulatory people who use doorknobs, towel bars, vanity tops, and the shower enclosure door to steady their trip to the bathroom. When wheelchair access is required, the environment should be able to be "taken apart" to accommodate turning radii and limited reach capacities. This is the concept behind the "friendly house," an adaptable unit created by Lewis Homes using National Council on the Aging criteria.

Considerate design is important with respect to reach capacities, grip strength, and visual acuity. If all environments were designed to be easily manipulated, everyone would benefit. Finally, furnishability is a key element in designing a space that is both flexible and accommodating. Key pieces of furniture should easily fit in a configuration that allows the room to be arranged in at least two different ways.

FIGURE 4.44 The dwelling units at the *de Kiekendief* housing project, in Almere-Stad, the Netherlands, are designed to accommodate aging in place. Bedfast residents can be moved through the front door by unlatching a second hinged partition that, when open, accommodates the width of a bed. This feature makes it possible to specify a residential entry door that is of normal width.

Unit Trends and Considerations

In general, the trend in Europe and the United States is toward larger units and away from small 200–250-square-foot rooms. Residents often prefer to have a separate bedroom conveniently located in relationship to the bathroom. Separating the living room from the bedroom allows greater privacy and makes it easier to accommodate overnight stays by family or special-duty nurses. A one-bedroom or studio alcove also allows residents to bring more furniture and personal items from their previous residence, easing their transition.

In Europe, the only major piece of furniture that must meet a separate performance standard is the bed. Disabled older residents who require assistance in transferring from bed to wheelchair can cause professional caregivers disabling back injuries. Beds for these residents are outfitted with hydraulic lifts and special features that facilitate transfer. Many European projects allow residents to replace wall and floor coverings as well as bring in light fixtures to prewired outlet boxes. These custom adjustments allow each unit to take on a unique personal expression. Antique light fixtures and chandeliers are visible features that greatly enhance the perception of the unit as personal and friendly. Adaptive strategies that allow residents to age in place as they become increasingly disabled involve accommodating a range of possible aids. In the **Küllengens Sjukhem**, in Hallsberg, Sweden, ceilings have been designed to receive a surface-mounted track which can be used to facilitate transfer from the bed to an adjacent wheelchair.

In general, however, the kitchen and bathroom are the most critical rooms in which adaptive changes are likely to occur. The flexibility of these rooms to adapt to equipment or assistive devices should be explored in early design stages.

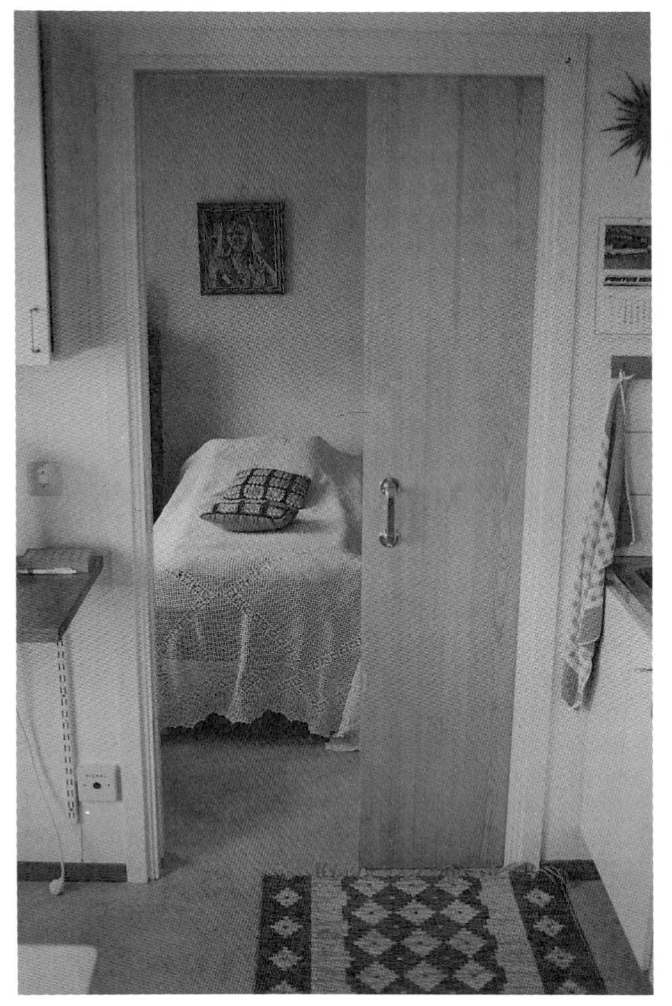

FIGURE 4.45 Danish units normally have wide sliding doors to bathrooms because they are easier for wheelchair and frail residents to open. They slide out of the way and by allowing portions of the adjacent corridor to be used, create a bigger space for transfers. Doors are either hung on the outside of the wall like this one, in barn-door style, or are recessed in the wall as pocket-doors.

Bathrooms Designed for Access

Bathrooms must be designed to facilitate a transfer from a wheelchair or walker. They should be large enough to allow attendants freedom of movement in assisting with toilet and shower transfers. These basic hygiene tasks should be performed safely and efficiently without placing the older person or the caregiver in jeopardy. Handicap standards often assume the person has adequate upper body strength for an independent transfer. Older people who need a wheelchair are often so frail, they have little independent ability to complete a transfer. Testing the size and configuration of space necessary for making a transfer and training older residents and caregivers to follow prudent safety guidelines are important.

Research focused on the needs of handicapped individuals (Steinfeld, 1987) has revealed the uniqueness of most handicap situations. On the other hand, standards are designed for the most generalizable situations and not necessarily for the specific disabilities and abilities of each individual. European bathrooms are often arranged to accommodate an open shower in one corner or on one edge of the bathroom. This type of design accommodates wheelchair access, making it easier for older residents and attendants to navigate within the room. Danish units almost universally specify a wide sliding pocket or barn door to the bathroom. This eliminates the problem of dealing with an inconvenient door swing and facilitates the use of larger width openings that can accommodate transfers more easily.

Kitchens Designed for Adaptability

Rosewood Estate, in Roseville, Minnesota, and **Rackleff House**, in Canby, Oregon, include full kitchens because of the options and choices those facilities provide. Residents in these facilities are supported by the normal features they would need to live independently in the community. When food storage and preparation spaces are not included, it is impossible for residents to exercise many independent behaviors. Specifying a larger unit but outfitting the setting with services that are prescribed based on need is the philosophy of the Danish and Swedish service house. In Sweden, small group homes for Alzheimer's victims frequently have a kitchenette. Stovetops are designed with timers and switches that permit residents to use them safely. Promoting and encouraging food preparation activities in European nursing homes can be traced back to early models that provided group kitchens with food lockers for residents. Newer buildings have increased this standard by providing small private kitchens in each dwelling unit.

The most common kitchen arrangements in European projects are open U- and L-shaped alcoves that accommodate seating. Kitchen tables located on one side of an L-shaped configuration are designed to be removed if the unit needs to be remodeled for wheelchair access. In the **Fyensgadecentret** project, in Ålborg, Denmark, freestanding kitchen cabinets are used to define the kitchen. Seven-foot-high cabinets allow a continuous ceiling between the living room and the kitchen, giving smaller units the perception of being more spacious.

In the **de Overloop** project, in Almere-Haven, the Netherlands, cabinets designed for kitchen utensils, clothing storage, and display separate the kitchen from the living room. The placement and design of cabinetry as freestanding furniture ensures privacy in the living room by blocking the view from the kitchen window to the corridor. Most service house kitchens are full size and many offer a range of appliances, including microwave ovens.

Design Details

Many projects utilize eight-foot window heights. These allow a greater amount of light to penetrate deeper into the unit. Several projects use glass wall panels between the living room and bedroom. These transparent planes of glass allow light to flow between rooms and give the impression of a larger dwelling unit. Residents who feel the need for greater privacy can close them off with curtains. At **Rosewood Estate**, in Roseville, Minnesota, a pair of French doors connects the bedroom and living room. The second door can be opened to accommodate wheelchair movement while the windows in the doors help to diffuse natural daylight.

Night-lights have been carefully located to better illuminate the path from the bedroom to the bathroom. This trip can be dangerous when made in the middle of the night. The Marriott **Brighton Garden** project, in Virginia Beach, has mounted a lighted translucent panel above the bathroom door to accommodate this. At the **Argyle**, in Colorado Springs, a wall-mounted luminescent panel next to the bathroom door can be switched on at the head of the resident's bed to light the pathway.

Emergency response systems are handled quite differently in Europe where response centers are often located at local service houses. Some projects have their own emergency call systems internally wired to a desk which is staffed day and night. Others have installed telephone-based systems which are monitored at another location. In the United States, **Rosewood Estate** uses a Lifeline system linked to mobile phones carried by attendants. The backup for the system is linked to local family members and then to a central control station in Boston. In the event a care attendant does not respond immediately, backup contacts are initiated.

Systems that provide daily reassurance of resident well-being are common in the United States and Europe. Many European systems have wired these reassurance systems into the refrigerator or toilet. When someone opens the refrigerator door or flushes the toilet in the morning, the mechanism automatically resets itself for the next day, assuring management that the resident is all right.

FIGURE 4.46 Clever ideas about how to increase storage space and create more flexible bathrooms are also common. In some service houses, larger wheelchair-accessible bathrooms are specified and filled with cabinetry which can be removed if a resident is confined to a wheelchair and needs extra space for maneuvering.

Summary

The dwelling unit should be flexible enough to meet the changing needs of the older person, considerate enough to increase mastery and control, and complete enough to deliver privacy and a symbolic sense of "home." Unit designs should be designed to make it easier for older people to:

1. Age in place through adaptable designs.
2. Manipulate and access appliances, fixtures, storage spaces, windows, and doors.
3. Preserve the privacy, sanctity, and autonomy of their unit.
4. Personalize and decorate their unit to suit their own needs and preferences.
5. Make the necessary bathroom transfers with assistance or by themselves.
6. Maintain the maximum amount of choice and independence within the setting.
7. Safely call for needed help and assistance when necessary.
8. Have an adequate amount of artificial and natural light.

13. RESIDENTIAL CHARACTER AND IMAGERY

Introduction

One of the major complaints about the appearance of institutions is that they look cold and uninviting. Most multistory hospitals, prisons, and nursing homes have flat roofs and utilize commercial finishes, fixtures, and surfaces. Design decisions are often based on maintenance concerns—how easy it is to fix or clean. Criteria like these are rarely a high priority in residential design. In fact, builders and real estate professionals frequently state that the affective feelings generated by a home's "curb appeal" motivate buyers to purchase. This transaction is affect-filled and involves measuring the buyer's personal perceptions of the house against projections of status, comfort, and fit.

Entering a nursing home rarely elicits positive feelings. In fact, moving out of one's lifelong house is a very difficult process. Marketing professionals often remark how traumatic the move is for many, because a lifetime of personal experiences and memories are stored in associations with the dwelling unit and furniture.

Residents want assurance they will feel as if they are at home. A building that looks large and commercial doesn't reassure the person contemplating a move. If a building appears cold, unappealing, and institutional, family members, visitors, and staff may assume behaviors consistent with that impression.

One of the most important design decisions is how to reduce the scale of the building so that it appears residential in character. Porches, overhangs, and dormer windows help to reduce the appearance of a larger-scale building from the street. Just as forced perspective in the streetscapes of Disneyland or Universal Studios give children the feeling of mastery over a large-scale environment, so can these design techniques help tame the scale of larger buildings.

Interiors are also subject to similar associative symbolism. Common rooms should be designed to be intimate in scale rather than overwhelming. Residents and visitors want to feel at home and not like they are visiting the local mall or high school gymnasium.

Finally, furnishings and accessories should be consistent with the interests and tastes of residents. Art that stresses emotional relationships between people can generate positive feelings and help recall positive past events. Accessories that rekindle past memories have meaning that goes beyond the shape, texture, and appearance of any item. Such approaches fulfill a resident's need for a comfortable and familiar residence.

Exterior Design Considerations

One of the most troublesome problems in creating housing that appears residential in scale is the need to accommodate enough units for efficient operation within a building envelope that doesn't appear overwhelming. The problem is further complicated when larger dwelling unit sizes are programmed on a relatively small urban site. Projects that deal successfully with this problem use various strategies to reduce the building's scale, making it seem more approachable.

One idea is to break a larger building into several smaller ones. The **Sunrise** retirement community of Bluemont Park, Virginia, constructed three L-shaped buildings of 50 units each, adjacent to one another, in place of a single 150-unit structure. The sloping site accommodates a below grade perimeter service corridor that connects all three buildings. This corridor allows laundry and food services to be centralized while giving the impression from the street above that each building is autonomous.

The design of **Woodside Place**, in Oakmont, Pennsylvania, also incorporated ideas about how to break a large building into smaller parts. The thirty dwelling units here are clustered in three small houses. From the outside the building mass appears as three houses adjacent to a fourth larger building, a pavilion which contains community rooms, food service, and administration.

Rosewood Estate also takes advantage of a sloping site and a tripartite massing scheme to reduce the perceived impact of the building from the street. The building rises two stories from the street in front while the sloping site allows it to be three stories at the rear, where it overlooks an adjacent lake.

Courtyard designs are also popular because they lend themselves to small sites allowing dense and compact configurations. The **Rackleff House**, in Canby, Oregon, is a good example of an attractively scaled courtyard form. The twenty-five units in this scheme are organized around two double-loaded and two single-loaded corridors that form the courtyard. This compact one-story configuration provides a substantial amount of natural daylight in corridors.

A final approach with great potential for suburban sites involves clustering several houses around a residential cul-de-sac. **Arizona Senior Homes**, in Mesa, Arizona, has done this by clustering six detached houses on two sides of a dead-end street. Each house, designed to accommodate ten residents, has its own living room and

FIGURE 4.47 The complex shape of *Woodside Place,* in Oakmont, Pennsylvania, which contains three residential bungalows and a large common area, has been characterized as reminiscent of a Shaker commune, where residential and meeting spaces were often in linked separate buildings.

warm-up kitchen. One larger dwelling contains a central kitchen and a multipurpose meeting room. This concept is interesting because it fits assisted living into a conventional suburban pattern. From the street the housing appears almost identical to adjacent single-family housing. This project demonstrates a viable solution to building assisted living housing in aging suburbs where larger-scale attached buildings are likely to be at odds with the surrounding context. A cul-de-sac street form allows a cluster pattern that minimizes trip distances and centralizes outdoor activities.

Roof, Hearth, and Front Door

What elements powerfully and convincingly reinforce residential imagery? Rybczynski (1989) suggests that the roof, hearth, and front door are the basic symbolic residential elements. From the exterior the most influential of the three is the roof, which symbolizes the notion of protection. Sloped roof forms have for centuries connoted shelter and security (Rapoport, 1969; Alexander, Ishikawa, and Silverstein, 1977). Sloped roof imagery is so pervasive in so many cultures that modernists choose the flat roof as a design idea precisely because it violates previously held conventions (MOCA, 1989).

Roof dormers can also reduce the perceived height of the building. Placing the roof eave line at the sill height of an upper-floor window allows a three-story building to be perceived as only two to two and a half stories in height, from the street below.

Another element that contains positive residential associations is the fireplace and the chimney. The hearth provides warmth, symbolic associations of togetherness, and visual fascination to users (Alexander, et al., 1977). The chimney, which has no place in the modern office building or hospital, reinforces the idea that this is housing and not a commercial or institutional building.

The friendliness of a dwelling is also affected by the entry sequence and the front door. An exterior entry porch should shelter visitors while they wait, provide a glimpse of the interior, and seem approachable through the use of plants and familiar details. This residential feature is often complicated by the desire to shelter residents and visitors as they disembark from their car. The **Sunrise** retirement community in Fairfax, Virginia has solved this conflict by designing a side entrance where a convenient, residentially scaled porte cochere is located. The formal front entry to the house is situated in the center of the building where the L-shaped wings of the building meet, which is most compositionally logical.

Residential Associations and Vernacular Traditions

Another element present in vernacular housing and recognized for its social and aesthetic influence is the front porch. In the memories of many older people, the front porch symbolizes a positive and friendly gesture to the neighborhood as well as functioning as an effective place to observe activities. Additionally, the scale of the porch and its attachment to the building often reduce the perceived mass of the structure from the street. Instead of a two- or three-story vertical wall, the porch, with its sloping roof, gives the building a more intimate, friendly look.

Landscape elements should create outdoor rooms. Complex building forms linking interior spaces physically and visually to exterior gardens, porches, and patios are particularly effective in adding variety and uniqueness to the environment. **Woodside Place**, in Oakmont, Pennsylvania, weaves together courtyards with residential housing blocks in a plan that daylights interior spaces and provides interesting views to surrounding site features.

The feeling of residential scale also depends on the design and placement of doors, windows, trim, and cabinet work. Commercial finishes and materials often confuse the residential imagery of the building. The technique of using finishes and materials consistent with industrial buildings to create new perceptions of a place is a powerful architectural idiom used by modernists like Richard Neutra and postmodernists like Frank Gehry. In the assisted living housing type, the architectural expression should reinforce the feeling of a residential environment through the selection of details, finishes, fixtures, wall and floor coverings, furnishings, and a spatial vocabulary that is consistent with perceptions of home. Flat roofs and automatic doors are features that have a place, purpose, and an architectural application, but do not foster associations consistent with a comfortable and familiar home environment.

Many inspiring ideas can come from the vernacular housing traditions of a region. Houses from the South, the West, the Southwest, the Midwest, and the East have different architectural heritages. They present a rich assortment of images and details that interpret needs and preferences of the people and forces of the environment in that particular region. Building materials, construction techniques, residential details, and housing forms are often based on the availability of particular materials and tradesmen.

Woodside Place, in Oakmont, Pennsylvania, took its inspiration from the Shaker commune, utilizing the bungalow and protective roof form of the rural barn. **Elder Homestead**, in Minnetonka, Minnesota, used details from nineteenth-century building construction catalogs inspired by the imagery of the rural farmstead. The **Captain Eldridge Congregate House** was designed to fit into a neighborhood of turn of the century Cape Cod houses in Hyannis, Massachusetts.

European housing projects often rely on adaptively remodeled buildings. The **Bouwacher** project, in Utrecht, the Netherlands, adaptively remodeled a nineteenth-century mansion house as the commons building for 111 housing units. A semicircular form was selected that capitalized on the diagonal siting of the old mansion house. The **St Bartolomeus Gasthuis**, in downtown Utrecht, which was was opened in 1367 and is thought to be the oldest continuously operated home for the country's aged, was remodeled by salvaging pieces from previous remodeling efforts in 1600, 1872, and 1927. Parts of the old building, including the entry door and foyer, use portions of the original building in the redesign.

Interior Design Considerations

One of the most compelling qualities of the Swedish group home is how effectively the scale of the dwelling promotes the perception of comfort and control for residents. The size of the dwelling is consistent with a large single-family house. Common spaces typically include a kitchen, living room, dining room, and garden terrace, which are similar in size and organization to a large residential family dwelling. Most group homes utilize eat-in kitchen arrangements to increase coziness and involve residents in food preparation activities.

The comfortable scale of the house is in stark contrast to the psychogeriatric hospitals they replaced. Although group homes cannot handle violent patients or highly agitated wanderers, the family setting seems to have a calming effect on residents and visitors. Local nursing homes in Sweden, as well as the Danish plegehem, are organized around the concept of decentralization. This creates a more intimate and manageable setting within a larger development. Local nursing homes and the group home have influenced one another as models for care. Recently remodeled local nursing homes like the **Vastersol** project, in Jönköping, Sweden, create a stronger sense of home by having residents and staff take meals together around a large table. In the **Solgård** nursing home, in Tranemo, Sweden, the small dining room for eight residents is flanked by an open serving kitchen which gives it a cozy residential feeling. The coffee, snacks, and meals served here and the use of incandescent lighting make it a comfortable place.

FIGURE 4.48 Danish service houses allow residents to hang their favorite light fixture and bring with them their favorite furniture. Residents normally replace wall and floor coverings when they move to a dwelling unit. This allows the resident to further personalize the place, making it their own.

Residentially Scaled Rooms

The design of appropriately sized rooms is important in preserving a sense of residential scale. The dining room is often troublesome, because it is usually the largest space in the building. Furnishings must accommodate wheelchair and walker access but should not sacrifice the feeling of residential scale. At the **Sunrise** retirement community on Mercer Island, Washington, the dining room, on the top floor, is designed as a series of alcoves. These are created by dormers carved out of the roof plan and the placement of a fireplace that divides the room. The layout maximizes the number of seating choices near the edge of the room and takes advantage of higher ceiling heights to introduce additional spatial variety. In most **Sunrise** facilities a schedule of two seatings minimizes the need for a larger dining room. This schedule allows the second seating to be for residents who experience eating difficulties.

Specific features common to effective images of residential housing are present in the **Captain Eldridge Congregate House**, in Hyannis, **Sunrise** retirement communities, and **Elder Homestead**, in Minnetonka, Minnesota. All three use an open stair to gain access to upper floors. A dramatic Victorian inspired split staircase is a *Sunrise* retirement home signature design element which is replicated in most of their facilities. The **Captain Eldridge Congregate House** places the stair in the middle of an open two-story atrium, flooding it with natural light and positioning it to overlook the dining area below. All three buildings utilize outdoor porch spaces to soften the look of the building from the street and provide convenient outdoor sitting areas for residents. The **Elder Homestead** and **Sunrise** projects have also employed enclosed and conditioned porches that can be used during colder weather.

Furnishings, Finishes, and Materials

Giving an interior the look of a residential environment is greatly dependent on furniture choices and finish decisions. Floor finishes, wall coverings, and casework selected solely on the basis of maintenance often look institutional and give residents the covert message that the setting is designed to anticipate incompetent behaviors. **Woodside Place**, in Oakmont, Pennsylvania, wanted the warmth and friendliness of wood floors in residential units. A vinyl product designed to replicate the grain pattern, appearance, and feel of wood was specified and looks genuine to the untrained eye. Details in each unit, inspired by Shaker designs, are used to create narrow plate shelves and pegs for the purpose of display and storage. Residents have collectibles on a three-inch ledge, and have hung clothing and hats from pegs, adding an element of personalization and variety to each room.

At the **Captain Eldridge Congregate House**, light oak floors and Cape Cod details have been combined with oak tables and chairs in the dining room to produce a friendly residential look. Most settings avoid the use of fluorescent lighting and use incandescent fixtures in combination with natural light. The top floor of the **Sunrise** retirement community in Fairfax, Virginia, is checkered with transparent skylights that open the room to light and views of surrounding trees. Twig and branch patterns in the winter and the green of summer foliage are visible. In recent **Sunrise** designs, upper-floor common rooms are opened to more natural light and views through the placement of octagonal sitting areas surrounded by windows at the edge of these spaces.

Giving the building the impression it is "normal" housing while still designing it to overcome ambulatory difficulties is a challenge. At **Sunrise**, some residents experience difficulty steadying themselves while walking. Instead of using a conventional handicap handrail on one side of the corridor, they have specified a 3.5-inch-wide chair rail ledge on both sides of the corridor. This design is far friendlier and also serves the purpose of steadying residents better than the bulkier and more stigmatized handrail. Searching for ways to make the setting more supportive without sacrificing its residential character is critical to success.

FIGURE 4.49 Artwork is used throughout *Sunrise* retirement community projects to set a mood and establish a feeling of emotional attachment between people and animals. They believe abstract graphics and modern art are less meaningful to older residents. *Sunrise* projects also use antique furniture and accessory pieces to provoke interest and conversations between residents and family members.

Artwork and Accessories

Seating designed to meet the flammability requirements, anthropometric needs, and the aesthetic expectations of residential character is often hard to find. In most cases, this furniture lends a monotonous sameness to each space, when specified throughout the building. Accessory pieces like tables, bookcases, desks, cabinets, and grandfather clocks can be used in conjunction with contract furnishings to add interest and variety. Antiques are often used with furnishing plans because they lend uniqueness and have a curiosity-provoking connection to the past. Residents have often owned similar items or have familiarity with the history of a furnishing piece.

Sunrise retirement communities deliberately display everyday antiques, from the time residents were growing up, in standing- and wall-mounted cases. Typically, collectibles from the turn of the century to the 1930s, like pocket knives, hair bows and combs, lace handkerchiefs, woodworking tools, kitchen items, and costume jewelry, spark conversations between staff, family members, visitors, and residents. These decorative items provoke and stimulate mental process. **Sunrise** projects have also selected artwork with a strong affective theme. Scenes portray intergenerational exchange between family members and the affection shared by children and pets. Instead of focusing on still life or abstract graphics, scenes that portray upbeat human experiences have been selected.

Woodside Place has focused on quilts and folk art, reinforcing the Shaker design theme reflected in architectural details. Quilts introduce a soft, colorful material that residents can touch. They are mounted for easy removal for replacement or cleaning.

Use of contemporary artwork has caused problems in some projects. In Sweden, one percent of the project's construction budget is set aside for artwork. This program, however, has led to artwork installations that are inconsistent with the interests and desires of older residents. The **Forsmannsenteret** project, in Tønsberg, Norway, had an acquisition program which became controversial. Artwork located in public spaces was acquired without input from older residents and users of the service house. The conflict between the tastes of art professionals and the interests of older visitors caused a major disruption in the community.

It is important for artwork to be evaluated on its therapeutic potential. Evoking memory, stimulating positive emotions, and sparking conversation with relatives and friends are important outcomes of a successful art program. This may narrow or violate the sense of freedom artists consider appropriate, but as long as assisted living is conceptualized as a therapeutic residential environment, the social and emotional impact of artwork on older residents must be considered.

Summary

Assisted living environments should appear residential rather than institutional. The scale, material selections, interior design treatments, furniture and accessory selections, and room dimensions should be consistent with residential environments rather than commercial or institutional settings. Sloping roof forms, the hearth, and the front porch are features that have traditionally been associated with residential housing. These elements evoke strong positive associations. Maintaining a residential scale while achieving appropriate operating economies is difficult to achieve. Breaking the building into several smaller houses is another effective approach.

Interior design treatments should use traditional residential materials and unique and unusual thought-provoking furniture items, rather than contract items picked out of a catalog. Finally, artwork must be viewed from a therapeutic perspective. In a setting designed to stimulate mental function and create friendships, the right artwork can create positive opportunities for interaction, while poor choices can make the setting appear alienating and unfamiliar.

14. ROOMS WITH A BEHAVIORAL PURPOSE

Introduction

It is a sad experience to visit a facility designed with strong intentions about how spaces are to be used, when those intentions have not been understood by management. Only one experience is worse, and that is visiting a setting where *no* explicit intentions for use or behavior were generated for social spaces. When rooms are designed without regard to how they will be used, meaningless furniture pieces often just fill the space. This leads to spaces that are "overlounged," giving residents and visitors a stilted and narrow impression of how common area spaces should be used. A behaviorally based architectural program allows the purposes associated with each space to be stated and inspected. This allows the types of activities planned for a space to be rehearsed. Understanding this informs design decisions so they can support particular ideas about use or embrace concerns about change and flexibility. It also imbues a room with meaning that can be reflected in furniture and accessory choices.

Room purpose and use may change in response to differing occupancy patterns that vary from the morning to the evening. Specific expectations for a space should be coupled with an understanding of how it can take on new or different roles when demands change. A behaviorally based architectural program should be designed like a trip itinerary or shooting script for a movie. It should be planned well enough to guide initial decisions but be flexible enough to accommodate changes, improvements, and adjustments that enhance usefulness.

Variety in room size and character is also an important attribute in maximizing flexibility. Rooms of similar size, which look the same, add nothing to the distinctiveness or the flexibility of the setting. Rooms can also elicit and support specific activities based on their shape, proportion, layout, and furnishing plan. Some common rooms should be designed to encourage informal activities, while others can be treated as special places to entertain guests or family. Common areas in the building should resemble the rooms of a normal house in variety, scale, purpose, and use. A parlor, family room, dining room, garden, breakfast nook, living room, and recreation room all have differing expectations for use. Residents may dress, behave, and act differently when rooms are designed to elicit different responses. Differences in appearance and expectation add variety and interest, making it more like a normal residential environment.

Rooms that Support Ideas About Use

Frank Lloyd Wright has been quoted as saying that architecture is "art with a purpose." Understanding how to establish and clarify purpose, how to accommodate changing purpose, and how to interpret purpose in an artful way is a fundamental concern of architecture. Rooms designed without underlying ideas about purpose are often boring, monotonous, and awkward. The possibilities for using a space are limited when there is little variety of expression to cue residents or staff. Understanding purpose can focus creativity and lead to interesting ideas about how rooms can support a range of activities. Sometimes the cue can be as simple as designing a furniture layout that accommodates a good off-site view or excellent exposure to sunlight in the morning for reading the paper. At other times, a sensitivity to the dynamics of group activities could be manifest in a room design that provides residents with an excuse to pass through while identifying potential partners for conversation at the same time.

Howell's evaluation (1980) of use patterns in housing for the elderly provides numerous examples of rooms where the lack of fit between management policy and the behavioral purposes established by the designer have led to unsatisfactory outcomes. Establishing how a room can support activity, stimulate social interaction, and provide the possibility of retreat doesn't occur by accident.

Buildings designed today often require a team of specialized professionals. A critical dimension of the success of a team effort is good communication. When interior designers who select furniture, woodwork details, and accessories do not influence the architects who are establishing the size and configuration of the room, the architecture may not be informed by an understanding of purpose and activity. When architects make drawings of rooms without a sense of how they are to be used, awkward spaces that do not fit the patterns of activity envisioned can result.

Designing intimately scaled residential rooms requires careful attention to these factors. Proof of the effectiveness of an integrative approach is obvious in classic houses by master architects such as Frank Lloyd Wright or Greene and Greene. In these settings, furniture fits into each room in memorable ways that support the activities and intentions. Purpose has been inspected and has become the basis for defining and shaping space.

FIGURE 4.50 The courtyard of the *Snostorps* service house, in Halmstad, Sweden, is a simple plan that involves two rows of single-story housing connected to a common building. An outdoor deck with a trellis cover expands the use of the large multipurpose room from which the foreground view is a colorful garden. In the background is the view of the adjacent valley.

One major difference between a custom home and a tract home is the degree to which the environment is fitted to the needs, possessions, and lifestyles of unique occupants. In *The Place of Houses* (1974), Charles Moore and his colleagues provide numerous examples of houses formed around the interests and activities of clients. In fact, one chapter of the book is devoted to a "kind of checklist" designed to uncover preferences and patterns that can lead a designer to creative insights about how to meet purpose.

The **Argyle**, in Colorado Springs, has used antique furnishings and built-in casework from the original 1899 building to display objects, frame the hearth, store books, and secure table game paraphernalia. The memorable quality of the furnishings and casework creates a comfortable place for conversation and reading.

Designing housing for nominal (undefined) clients rather than substantive (identified) clients is often the rationale for generalizing architectural treatments. If the interests and preferences of soon to be residents cannot be identified, how can a personal architectural response be provided? Activities that offer choices, stimulate social exchange, or facilitate the observation of activity are purposeful enough to suggest a design response. In the **Snostorps** service house, in Halmstad, Sweden, two rows of four units each create a U-shaped configuration around a garden. The service house links the two residential wings facing the southeasterly view. The sloped site creates a platform that overlooks the valley. The architects layered the service house with a garden, outdoor patio, and glass wall that invites natural light in the morning. Taking morning coffee here is a delightful experience in the summer and winter.

Spaces that Anticipate and Support Behaviors

Behaviorally based architectural programs and design behavior hypotheses are useful techniques to elicit and specify the intentions of spaces and their relationships to one another. When these techniques are used properly, they create a shared image of the setting that promotes communication between the design team, the administrative staff, and residents. This is particularly important in assisted living where therapeutic considerations should influence design decisions.

Postoccupancy evaluations of these settings often focus on minor considerations which make a major difference in how well a space works. For example, an outdoor space may be found to be unpopular for a number of reasons that individually or collectively discourage its use. A convenient door may not be available. A higher than normal sill height may limit wheelchair access. A lack of nearby storage may limit access to recreational equipment. Electrical power and lighting availability may be insufficient. Door-locking policies implemented for security may also preclude access. Uncovering impediments and barriers to use allows purpose to be better served rather than impeded. Another advantage of a behaviorally based architectural program is the challenge it presents for management to think about how spaces can best be used and adapted as a population ages. A space can be proactive and suggest how it can be used, or it can deliver an ambivalent message. The adage that a "multipurpose room is a no-purpose room" summarizes this concern. It postulates that space trying to be all things to all people can end up being of no use to anyone. On the other hand, conceptualizations of purpose must embrace change. Spaces must be flexible enough to be adapted or remodeled for other uses. A strategy that rehearses possible use patterns and takes a flexible but thoughtful approach to future possibilities should be advocated.

What Establishes Purpose?

The most powerful influences in communicating purpose are the furnishings, equipment, and finishes in a room. The character of the interior treatment sets expectations for behavior by facilitating and supporting various activities. Rooms like the greenhouse at **Rosewood Estate** and the breakfast alcove at the **Captain Eldridge Congregate House** support informal use. Others, like the Rosewood Parlor at **Rosewood Estate**, or the dining room in the Mercer Island **Sunrise** retirement community in Seattle appear more formal. Residents respond to these settings by thinking of them and using them differently. Equipment placed in a room, be it a piano or an exercise treadmill, provides direct cues about the purpose of a space. Furniture placed around a conversation circle connotes group socializing just like a single easy chair adjacent to task lighting, a view, and a newspaper rack connotes a solitary form of use.

Certain furniture pieces also have more flexibility and support a broader range of activities. Tables, for example, facilitate card games, allow intimate conversations, make it easier for wheelchair patients to participate, and provide a support surface for drinks, snacks, and reading material. Tables are more flexible than a couch, two chairs, and a coffee table. Lounge furniture may be harder to enter and exit, more difficult to arrange in close proximity for the hearing impaired, and less flexible for alternative uses. Each furniture item has its place, but an overdependence on one type limits activity. For example, a breakfront storage cabinet can provide a place for game storage, books, and a house scrapbook, making it possible to activate an adjacent space. A piano in a small music room or a hard surface table and a sink in a crafts room suggest and facilitate certain activities.

Behavioral settings like the Dutch "brown café" elicit in the minds of residents behaviors, attitudes, and interaction patterns appropriate to that type of social environment. At **Woodside Place**, a small country kitchen was specified and has become popular with residents and family members. This may be because the kitchen table is a traditional setting for family interaction. In this case, eating is an activity dementia victims and their families can still enjoy together.

Furnishing decisions can reinforce therapeutic activities like reading when materials are displayed in racks and arrayed to entice residents to read them. Magazines and newspapers give those with shorter attention spans something to see and browse through. Large-type books are a special resource for the visually impaired resident. Additionally, increasing creature comforts with an easy chair or access to a cup of coffee or tea adds another dimension to the experience. Encouraging reading may also mean isolating it from the sonic and visual intrusion of a television set. Designing rooms that are familiar and comfortable makes it easier for residents to imagine how they can be used and create lively norms for future use.

Staff Impacts on Use

Staff and management also play a key role. Because furniture in room layouts are like props on a stage set, management must take a proactive role in "scripting" behaviors and activities that occur there. Encouraging someone to read in a room designed to support this activity gives others an impression the room can be used in this way. In some cases, management may need to prime the pump by getting a group together for a daily card game or by starting a puzzle and occasionally helping to complete it.

Group homes in Sweden have done this masterfully by training attendants to encourage residents to carry out normal daily activities such as folding laundry, setting the table, and cleaning their room because it stimulates mental and physical function. Activity begets activity. When spaces are filled with residents and visitors, they become more interesting places to visit and suggest different ways of using the spaces.

In this regard, it is reminiscent of the counterintuitive findings of William Whyte's 1980 study of vest pocket parks in New York City. He found that during peak periods of use park visitors rated the busiest and most crowded plazas as the quietest and most relaxing. Stimulating activity can give a setting vitality and a sense of engagement that is impossible to produce through other means. The opposite also occurs. When spaces are not used, they seem uninteresting. When a space does not support activities, this lack of use can create a downward spiral that discourages future use.

Finally, common rooms should encourage diversity of use, allowing residents to engage in different activities at the same time in the same room. Environments for the frail often rely on a "one size fits all" activity program. These are often geared to large events as opposed to individual programs that seek the involvement of smaller groups.

FIGURE 4.51 A brown café was created as a meeting place in the *de Drie Hoven* facility, in Amsterdam. It exerts a powerful presence because it utilizes the image and character of the neighborhood bar, which has been a neighborhood social institution in Dutch history for centuries.

Summary

Each housing project should be designed with a thoughtful assessment of how common spaces are likely to be used and how they will fit the therapeutic and social program envisioned. Staff input to a behaviorally based architectural program can focus creativity about furniture selections and layouts. Variety in room size and character maximizes flexibility and adds to the distinctive character of the environment. Rooms should be designed to support activities that increase communications between family members and engage residents in an active and satisfying collection of events.

Rooms can provide a number of cues about how they can be used for a variety of activities. Sometimes they can rely on familiar imagery such as the country kitchen at **Woodside Place** to cue behavior. Finally, flexibility and adaptability should be considered for the future. A billiards room designed as a "men's only" room may go unused when the number of male residents decreases. Environments change in response to the needs, preferences, and interests of residents, as well as the needs and initiatives of staff. Flexibility and thoughtful purposefulness are necessary criteria for success.

15. UNIT CLUSTERS

Introduction

One of the greatest benefits associated with an age-segregated group living arrangement is the increased opportunity it provides for friendship formation and social exchange. Older people living alone in the community can be subject to loneliness and isolation. In some cases this leads to depression and a lack of interest in health and nutrition. Design ideas that aid the development of informal helping relationships should be a priority.

Clustering units in small groups appears to greatly benefit social interaction. Propinquity within buildings with high concentrations of older people has proven to be a positive influence on social exchange and friendship formation (Rosow, 1967). Unit clusters let the architecture of the building create opportunities for residents to get to know one another better. These informal contacts often lead to deeper, more substantial helping relationships that decrease reliance on formal care provision. When people help one another, they gain a sense of satisfaction which increases self-esteem. Creating networks of interdependence between residents and between residents and family members brings a human dimension to the setting that is not present when caregiving is solely dependent on paid professionals.

Another benefit of small clusters is the opportunity they present to enhance architectural expression. Clusters allow the building to be massed into smaller, more intimate configurations. Nursing homes and hospitals are associated with contiguous, monolithic, and centralized schemes. A final benefit of clustering is that it allows care administration and service delivery to be decentralized. Creating several small dining rooms adjacent to unit clusters instead of a single large dining room increases intimacy. Smaller dining rooms located throughout the building make it easier for less ambulatory residents to reach the dining room on their own. Taking a meal in a smaller, more intimate dining room also increases the feeling of "family" solidarity. Projects smaller than twenty-five units benefit less from these techniques. For example, the **Captain Eldridge Congregate House** and **Rackleff House**, in Canby, Oregon, are relatively small-scale endeavors that have used floors and courtyards to separate and cluster units.

FIGURE 4.52 Danish nursing homes like *Nybodergaården,* in Copenhagen, provide several choices for residents to take meals. Residents can eat in their room, at a first-floor café, or family style at this small table located on one end of each floor. The table overlooks the entry forecourt.

Decentralized Schemes

Swedish local nursing homes and small group homes have based cluster sizes on the number of residents who can comfortably sit around a single large dining room table. This enhances communication and the ability to motivate residents to engage in independent activities and behaviors. When care administration is decentralized, residents have familiar attendants assigned to their needs. When friendships between staff and residents occur, they promote humor and good-natured feelings. This in turn can lead to better communication with family members, further increasing family participation in the lives of older residents.

Decentralized management is also a powerful organizational technique. It can nurture the unique qualities of each unit cluster, allowing them to develop characteristics, values, and norms that represent the people living there. The best examples of this idea occur in Danish nursing homes where systems of care, decentralized at the ward level (fifteen to twenty units), result in an increase in cluster unit uniqueness. Caregiving philosophies in these settings blend the personalities of the staff with the preferences and interests of residents.

At **Dronning Anne-Marie**, in Copenhagen, the equivalent of the "quality circle" in production management has emerged. Innovation flourishes and the competition between staff in separate clusters enhances care delivery. Another interesting by-product of these cluster differences is the change that occurs in the physical environment. Originally, each ward was identical in plan. Modifications in furniture, equipment, and artwork as well as remodeling changes in kitchen and office spaces over time have made each cluster unique. This demonstrates the strength of the approach, which allows each setting to develop its own individual character, similar to what might be expected over time in a subdivision of houses constructed by one builder. Residents and staff have transformed each cluster in a way that reflects their personalities and preferences.

Corridor Clusters

Large buildings can be imposing and overwhelming to the older frail person entering a group living arrangement. In general, the housing form should provide a replacement for neighborly contact. One effective way to do this is to cluster small groups of units around a common entry or small living room.

In a double-loaded corridor portion of the **de Overloop** project, in Almere-Haven, the Netherlands, four unit entries are clustered around alcoves. A corner window at each unit entry creates a visual connection between the four units and the corridor alcove. In other projects, ceiling heights, changes in floor materials, ambient light level increases, and personalized additions to unit entries have been employed to give these alcoves greater visibility and emphasis.

Another common approach has been to cluster six to eight units around an interior living room. Typically, the space is laid out to encourage residents to socialize. The **Sunrise** retirement community in Fairfax, Virginia, has clustered six units around each of four lounges on the upper two floors. These spaces are interesting because they are sized and furnished to resemble a "nor-

FIGURE 4.53 The *Vickelbygården* nursing home, in Skärblacka, Sweden, is organized around several courtyards that contain raised planter beds. Residents tend the gardens and everyone benefits from its color and fragrance.

mal" living room. Items like desks, bookcases, and tables are placed against walls, while couch and chair arrangements are centered around a television or a conversation circle.

In some buildings like **Rosewood Estate**, these lounge spaces are too large. They result in furniture arrangements that "float" in the middle of the room. This placement makes them appear institutional and gives the room an anonymous scale that takes away its residential feeling. The **Elder Homestead** project, in Minnetonka, Minnesota, has created an interesting solution by clustering four dwelling units around a shared parlor space. Dutch doors and double-hung windows are used to create an identity for each dwelling unit. The resulting cluster parlor relates well to each unit but feels distinctly communal.

The **Old People's Home and Health Center**, in Oitti, Finland, has created a decentralized lounge that serves the ten units that surround it. In this design the lounge space has been topped with a dramatic pyramidal skylight clearly visible from outside the building. A corridor links four of these lounge areas and treats each unit cluster as a "village." The large expanse of skylight glass lets natural light flood the space and plants flourish, giving it the feeling of an atrium or greenhouse rather than an enclosed room. A dining room located on one side allows residents sitting in each lounge to see outside.

Another approach has been to centralize community spaces by placing them between unit clusters. This reduces the corridor length, thus increasing accessibility from the dwelling units. At the **Solbacken** group home, in Okero, Sweden, a simple rectangular building with the kitchen, dining room, living room, and administrative area in the center of the building mass has two short corridors leading to four units on each side. The **Strand** group home, in Arvika, Sweden, uses an L-shaped building configuration to accomplish the same goal. The change in corridor direction foreshortens the perceived length of the building from the inside. Furthermore, the central outside space created by the building configuration is visible from the units and is protected from the wind. The advantage of these schemes is that they resemble a house in scale and organization.

Another similar approach is the atrium house, in Uplands Väsby, Sweden. Six units form a U shape around the atrium, which is open to a view at one end. The center area is used for shared activities. Shared plants grow alongside individual furniture items residents have brought into the space. The hybrid exterior-interior quality of the atrium works well as a communal setting for neighborly social contact.

Courtyard Schemes

Courtyard schemes are the most intriguing. One reason is that they make views and direct access to outdoor spaces an important part of the solution. Furthermore, courtyard arrangements are often easier to attach to one another, creating a system of attached housing rather than separate objects.

The **Vickelbygården** nursing home, in Skärblacka, Sweden, is a courtyard scheme that clusters fifteen units around a courtyard. The living room is located near the center of the cluster, while a dining room is shared between two adjacent clusters. The exterior courtyard is left open at one corner to give it access to the adjacent grounds.

The **Søreidtunet**, near Bergen, Norway, is a row housing scheme linked by a paved pedestrian street. It involves parallel rows of units placed against a gently sloping hillside. At the base of the hill is a service house, which opens onto a mixed-use commercial plaza that includes a post office and health center for the surrounding community. **Søreidtunet** utilizes an orthogonal network of pedestrian walkways inspired by the old Hanseatic settlement in the core of old Bergen, where narrow pedestrian walkways link long thin buildings to the street (Det Hanseatiske Museum, 1982). The spirit of the pedestrian linkage is present here in the organizational network of walking streets that link the houses to one another and to services below.

Courtyard schemes are represented best by the Dutch höfjes, which create a homogeneous community by virtue of their resident selection process and their segregated and protected form. The **Höfje van Staats**, in Haarlem, is a twenty-unit urban courtyard scheme. The separation from the city coupled with a convenient urban location offers a setting that is serene, safe, and beautiful as well as highly engaged with city street life. These settings, first created in the sixteenth century, are good examples of sociopetal housing forms, which have functioned well for residents for more than 400 years.

Another old but still attractive courtyard organization schema is the eighteenth-century Norwegian tun. **Lårdal**, in Høydalsmo, Norway, is a small rural service house that expresses the spirit of the tun. The original tun contained farm animals and was protected by fences, gates, and the building perimeter. The tun, as interpreted in this housing project, is the space between buildings where residents congregate.

A final cluster idea involves breaking a contiguous building into several separate unit clusters around an elevator and single stairs. These schemes have only one stair exit for a multistory building, and are not legally acceptable in the United States. These projects center around

FIGURE 4.54 *Höfje van Staats,* in Haarlem, the Netherlands, is an eighteenth-century social housing experiment. Twenty units surround a beautifully scaled courtyard, which contains a single large tree in the center. A paved perimeter around the courtyard serves as an entrance sidewalk to each unit and a place for movable seating and resident socializing. It has been beautifully restored and is a popular tourist attraction.

the circulation stair and elevator and typically share a common wall with adjacent clusters. The outcome is a contiguous building form that allows units to be served by point towers rather than by long, anonymous double-loaded corridors that lead to exits at each end of the building. **Tornhuset** in Göteborg, **Kvarteret Karl XI** service house, in Halmstad, and the **T-1** service house in Linköping, all report this configuration is very successful in accommodating their Swedish residents in vertical clusters around stairwells. This building form also allows units to be built with outside exposures on two sides, facilitating light from two directions and cross ventilation. Furthermore, units share an entrance at the base of the tower that gives each vertical stack of units identity within the whole.

Summary

Unit clusters are an important fundamental social building block that enhances the friendliness of a setting and counters the anonymous, contiguous, and monolithic nature of a large building form. Small clusters of units increase opportunities for friendship formation, reduce the scale of the setting to a smaller intimate residential proportion, give the building greater variety and spatial complexity, and facilitate a personal, decentralized management approach that can involve residents in self-governance. Unit clusters that utilize courtyard configurations are particularly well suited to relate indoor and outdoor spaces to units and common dining rooms.

10 CASE STUDIES

5

INTRODUCTION

One of the most effective ways to learn about buildings is through case study examples. In fact, architecture is often evaluated, understood, and taught by studying precedents that reveal previously explored ideas. Case studies offer a specific solution to a particular program on a unique site. Although their generalizability may be limited, they provide comprehensive, fully executed examples of projects. Each solution is unique in its response, given the multitude of potential directions and considerations available. Case studies can show how differing contexts with differing programs lead to building forms consistent with the nature of the problem. In this sense, they bring variety and diversity to our understanding of the possibilities for design response.

Buildings can be metaphorically thought of like people. One gets to know them well by spending time with them. Visiting a building is like meeting a person for the first time. The personality of the building is ingrained in its program of spaces and the purposes established for the setting. Its physical attractiveness is more judgmental, but involves a sense of appreciation for the detail and composition of its form.

This chapter reviews my "conversations" with ten buildings. Visiting them, experiencing them, and asking questions to users and staff about them has left me with a set of impressions. This chapter chronicles those impressions and communicates how the qualities and characteristics of each setting led me to feel about the architectural ideas that were being tested.

I am often frustrated by housing case studies that deal only with the physical nature of the architecture without considering the people who live there and the transactional relationship between people and environment. This added information is particularly important in housing designed for older people, because as people become frail, they are more dependent on the environment to accommodate their needs. In old age, housing can become a source of comfort and delight or a prison that entraps and stifles. Knowing who lives there and how the environment supports their needs helps us to better imagine the life of the place and how well it supports and facilitates the experience of living there. These basic attributes of a setting are what nurtures the spirit of the inhabitant and leads to a more fulfilling, richer architecture.

The case studies begin with some selected characteristics of the place and the residents. These allow you to understand each building in comparison with one another and with other places you have experienced. Hopefully, this information will allow each example to be more useful and meaningful. The case studies are annotated with statements that chronicle my impressions of noteworthy attributes. These "lessons learned" from each building comprise aspects I would study if I had only a short time to learn all I could from each. These notes are weighted toward positive ideas, qualities, and characteristics worthy of further exploration.

CASE STUDIES

These projects have been selected from the 125 site visited American and European projects. A selected list of the 100 best projects appear in Appendix A. They are listed in rank order sequence from the largest to the smallest. One project from each country has been included with the exception of the Netherlands (two) and the United States (four). These case studies are far from perfect examples, but each represents a unique point of view about architecture and caregiving.

1. **Jan van der Ploeg, Rotterdam, the Netherlands**
2. **Rosewood Estate, Roseville, Minnesota**
3. **Bergzicht, Breda, the Netherlands**
4. **Kuuselan Palvelukoti, Tampere, Finland**
5. **Nybodergaården, Copenhagen, Denmark**
6. **Sunrise, Fairfax, Virginia**
7. **Woodside Place, Oakmont, Pennsylvania**
8. **Captain Eldridge Congregate House, Hyannis, Massachusetts**
9. **Lesjatun, Lesja, Norway**
10. **Hasselknuten, Stenungsund, Sweden.**

CROSS-PROJECT COMPARISONS

Table 5.1 displays sixteen selected characteristics of each building and its resident population to aid comprehension and understanding of the wide diversity represented in the ten case studies. Although all projects are designed to provide supportive services to residents, they vary with regard to the nature of that support and the current needs of residents. The two Dutch "point of support" projects contain some of the youngest residents. These projects are designed to allow residents to "age in place" and contain spaces that can be transformed for intensive service provision when required in the future.

Projects also vary in the extent to which health care services are provided. Only two are designed as health care (nursing home) environments. However, it should be noted that northern European projects are more committed to keeping people independent in normal residential housing for as long as possible. The Hasselknuten group home for dementia victims and the Kuuselan Palvelukoti service house are designed with the intention of accommodating individuals with severe mental disabilities until death.

Physical Characteristics

Projects are located in a variety of contexts, but the majority are in suburbs of larger cities. They vary in size from six to seventy-nine units, with an average size of forty-two. The majority of the dwelling unit stock represented in these ten examples is one-bedroom. However, European one-bedroom units are smaller than their United States counterparts, with an average size of 500 square feet. Five of the case studies are atrium designs. Two of these have gardens on the ground floor while the other three contain activity spaces. Two projects are L-shaped, two others are rectangular buildings, and one is a complex courtyard form. Seven different housing/service types are represented.

Resident Characteristics

The average age of residents by project varies from a low of 70 to a high of 85.5, while the average across all projects approximates 80. The Bergzicht project has the highest percentage of couples. This project includes sixteen units of "lean-to housing," which is traditionally oriented toward younger retirees. Three other projects have between 5 and 10 percent couples. Measuring resident competency is difficult in European projects because service provision is handled by district homemaker and home health nursing organizations that are often unaffiliated with project management personnel.

Assistance with bathing is a good indicator of personal care need. Bathing assistance ranged from 18 to 100 percent across all projects, with an average around 50 percent. The range for memory-impaired and confused residents was from 0 to 100 percent, due to the inability of some buildings to contain and safeguard mentally impaired residents. The Woodside and the Hasselknuten case studies are designed only for dementia victims and therefore a preliminary diagnosis of the disease is necessary for admission. The range of service provision for toileting assistance and incontinence care is also great. In projects with younger populations or where assistance is ordered on an occasional basis, the number of residents with these problems is relatively low. All projects are designed to accommodate this eventuality. In one case, the size of the project, its small-town location, and the policies of the local service agency have made it difficult to provide night care and toileting assistance on a twenty-four-hour basis.

Table 5.1 CASE STUDY BUILDINGS AND RESIDENT CHARACTERISTICS

	State/ Country	# of Units	Contxt S/U/R Suburb Urban Rural	Type	Bldg Shape	Unit Mix	Unit Size (ft^{-2})	**HCS/ PCS	Avg Age	# of Residents	# of Couples	% Bath Help	% Conf. used	% Incontinent	% Toilet Help	# of Stories
Jan van der Ploeg	Europe (NE)	79	Urban	Steunpunt	Atrium (garden)	67 1-Bdr 12 2-Bdr	425	PCS	77.0	84	5	18%	2%	NA	NA	4/7
Rosewood Estate	USA (MN)	68	Suburban	Apt w/ Service	Rectangular	48 1-Bdr 20 Stu	550	PCS	82.5	69	4	50%	6%	20%	30%	2/3
Bergzicht	Europe (NE)	58	Suburban	Steunpunt	Enclosed Street	32 1-Bdr 26 2-Bdr	650	PCS	70.0	78	20	NA	3%	NA	NA	4
Kuuselan Palvelukoti	Europe (FI)	51	Suburban	Service House	Atrium (garden)	43 1-Bdr 8 Stu	500	HCS/PCS	76.0	57	6	10%	20%	5%	14%	4
Nyboder-gaården	Europe (DK)	54	Urban	Nursing Home	Atrium (activity)	48 Stu 6-Stu (large)	260	HCS	82.0	54	0	83%	35%	50%	83%	3/BSMT
Sunrise	USA (VA)	47	Suburban	Home/ Aged	L	12 1-Bdr 35 Stu	275	PCS	82.0	60	0	100%	40%	25%	25%	3/BSMT
Woodside Place	USA (PA)	30	Suburban	Group Home	Courtyard	24 S-Occ 6 D-Occ	180	PCS	80.8	36	0	92%	100%	28%	69%	1
Captain Eldridge	USA (MA)	18	Small Town	Congregate House	Atrium (activity)	2 1-Bdr 16 Stu	275	PCS	83.2	19	2	32%	21%	16%	0%	2
Lesjatun	Europe (NO)	15	Small Town	Service House	Cluster	15 1-Bdr	490	PCS	78.0	15	0	53%	0%	27%	0%	1/2
Hassel-knuten	Europe (SW)	6	Suburban	Group Home	L	6 Stu	340	HCS/PCS	85.5	6	0	100%	100%	83%	83%	1/*

* Located within a 2-story family housing block.

**HCS = Health Care Service

PCS = Personal Care Service

JAN VAN DER PLOEG, Rotterdam, Netherlands 1

Architect: EGM Architecten bv
Dordrecht, the Netherlands

Sponsor: Stads Vernieuwing
Rotterdam, the Netherlands

Building Characteristics:

a.	# of units	79
b.	# of stories	4 and 7
c.	Context	urban
d.	Housing type	steunpunt project
e.	Building parti/shape	atrium
f.	Unit mixture	67 1-bdr 12 2-bdr
g.	Size of most common unit (average)	425 sq ft
h.	Community facilities	meals program
i.	Community-accessible restaurant	yes
j.	Year opened	1988

Residents Characteristics:

a.	Average age	77 years
b.	# of people	84
c.	# couples/men/women	5 couples 10 single men 64 single women
d.	% Bathing help	18%
e.	% Toileting help	NA
f.	% Incontinent	NA
g.	% Wheelchair-bound	2%
h.	% Cognitively impaired	2%

FIGURE 5.1 The atrium at the *Jan van der Ploeg* project, in Rotterdam, the Netherlands, has an enclosed landscaped courtyard. An umbrella table near the entry and the multi-purpose room are popular places for residents to meet. Ventilators on the roof controlled by a thermostatic switch open to exhaust hot air.

Summary Description

This seventy-nine-unit garden atrium conforms to an unusual site in an urban renewal district of Rotterdam. The four-story building has a triangular-shaped plan and is covered by a glass atrium with a curved seven-story housing block at one end. This is a steunpunt, or point of support, project, which provides services such as nursing, home help, and meals to housing residents as well as to older people living in the surrounding neighborhood. The project was reviewed by a local committee of neighborhood residents that provided significant input to the architectural program and met periodically during the design process to discuss refinements.

Volunteers from the project take responsibility for managing the second-floor grocery store and assist a local social service agency in managing a luncheon program located on the ground floor that serves twenty-five neighborhood residents. The management philosophy stresses self-maintenance and interdependence, and residents are encouraged to help one another. The large atrium provides a protected place for activities and plants. The single-loaded balcony corridors that link units are curved and wide enough to accommodate a small table and chairs that extend each unit's territory into the atrium. Flowers in planter boxes attached to balcony rails give the atrium a lively, colorful gardenlike character.

Project Features

1. Atrium

The four-story garden atrium provides excellent daylighting to the center of the project. A paved portion of the atrium near the entry accommodates chairs and umbrella tables, making it a pleasant and inviting place for residents to meet and pass the time of day. The atrium plan is shaped like a raindrop with a narrow triangular garden at one end and a larger semicircular patio space at the other end. Dwelling units on all sides of the atrium are connected by a single-loaded balcony corridor. The balcony width is four feet, but a curved edge provides an additional 1.5 feet at each entry door to accommodate the placement of a small table and up to two chairs for viewing. Red geraniums mounted in specially designed flower boxes are attached randomly to the top of the handrail throughout the atrium. The atrium is ventilated in the summer by twelve large panels mounted near the center of the glass skylight. Thermostatically controlled dampers on the second floor supply makeup air when roof ventilators are open. The atrium is lively, filled with the sounds of activities generated by residents.

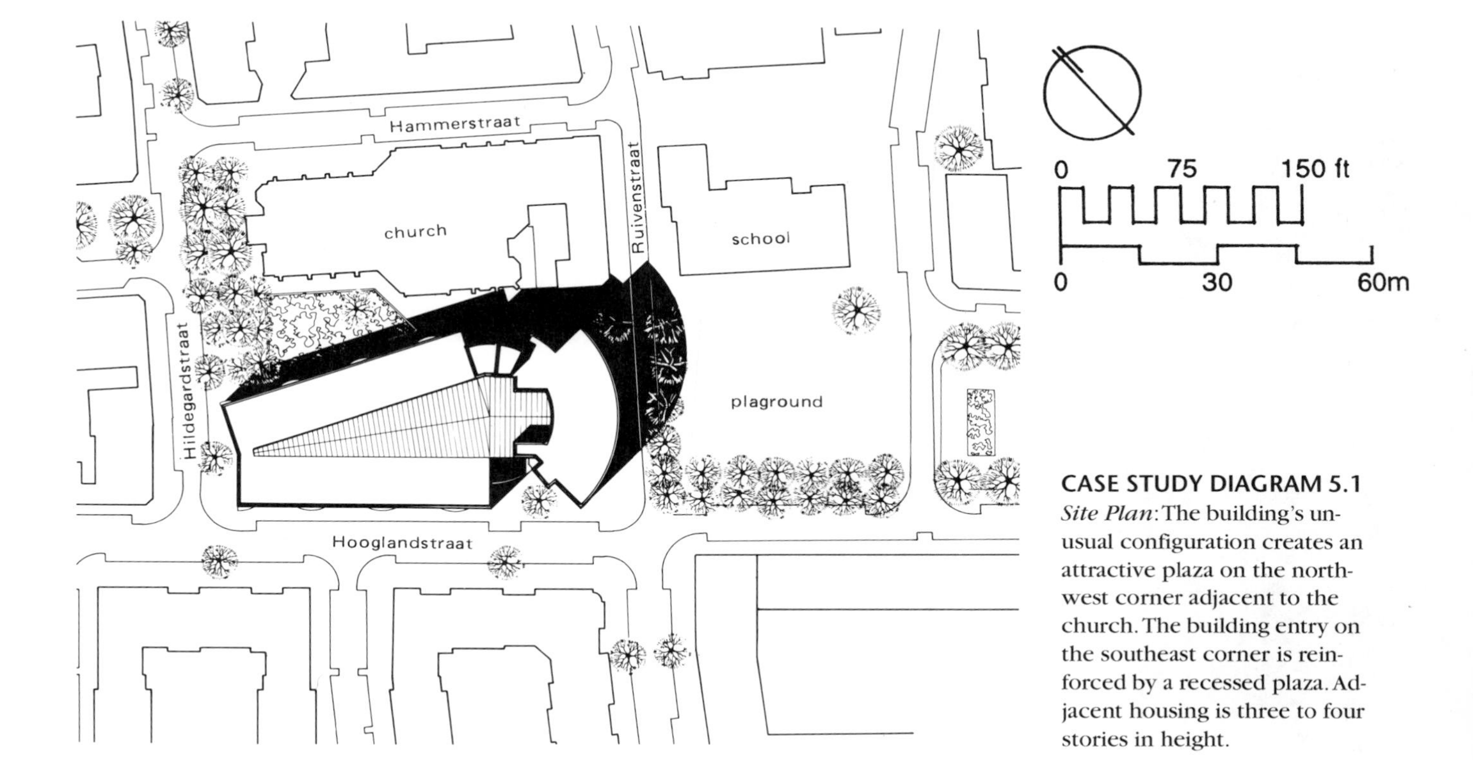

CASE STUDY DIAGRAM 5.1
Site Plan: The building's unusual configuration creates an attractive plaza on the northwest corner adjacent to the church. The building entry on the southeast corner is reinforced by a recessed plaza. Adjacent housing is three to four stories in height.

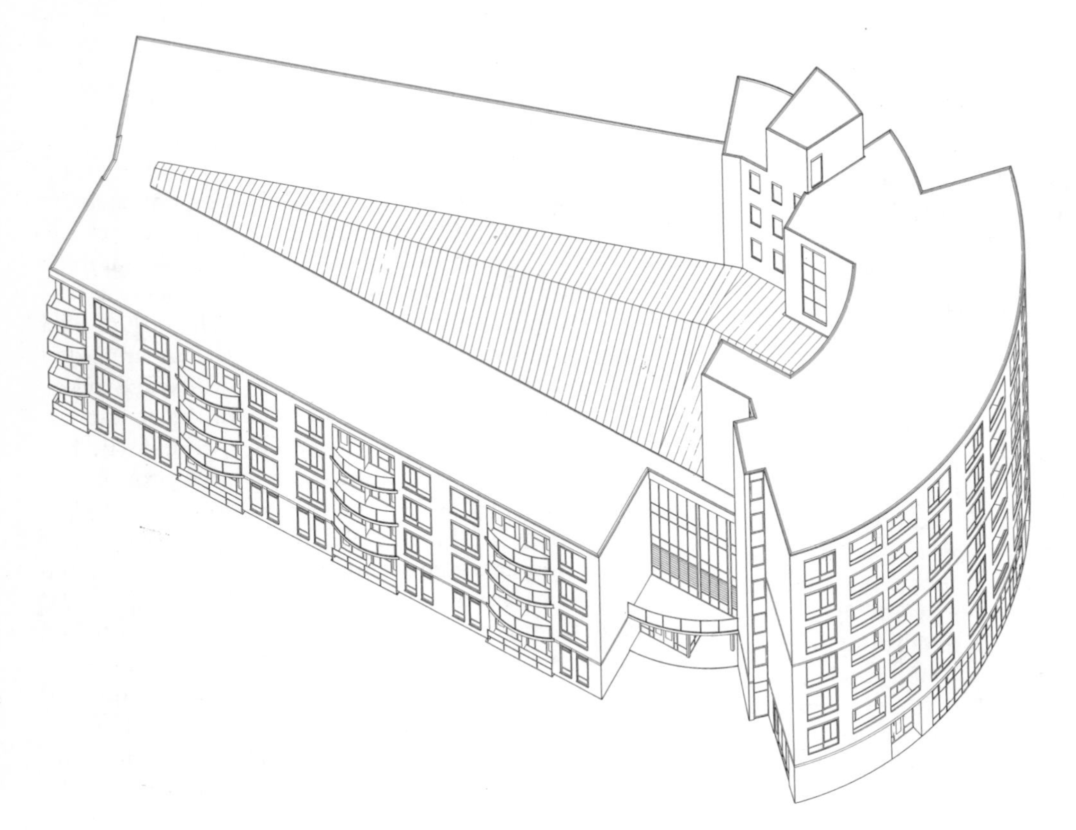

CASE STUDY DIAGRAM 5.2
Axonometric: The building's four-story V-shaped mass surrounds a garden atrium. The atrium is terminated by a curved seven-story housing block on the east.

2. Point of Support Project

Jan van der Ploeg offers services to project residents and older people in the neighborhood. These services include home nursing help, podiatry, and speech and occupational therapy. Personal care services provided keep residents independent and encourage aging in place. A multipurpose room and warm-up kitchen on the first floor provide meals for some residents of the building and about twenty-five older people from the surrounding neighborhood. Residents are not required to take meals. In the afternoon, a bar provides beer, wine, soft drinks, coffee, and snacks for those who stay and play cards or socialize. Resident volunteers do much of the work to support these activities.

3. Site

The site is located in an urban renewal district. In an area of mixed land use, it is adjacent to shops, schools, a large park, and a church. Designed to be self-contained, the project complements the surrounding context.

FIGURE 5.2 The single-loaded balcony corridors that link units are shaped to accommodate a table and chair adjacent to each unit entry. Red geraniums in planters attached to the balcony rail add color and delight to the balcony walls of the atrium.

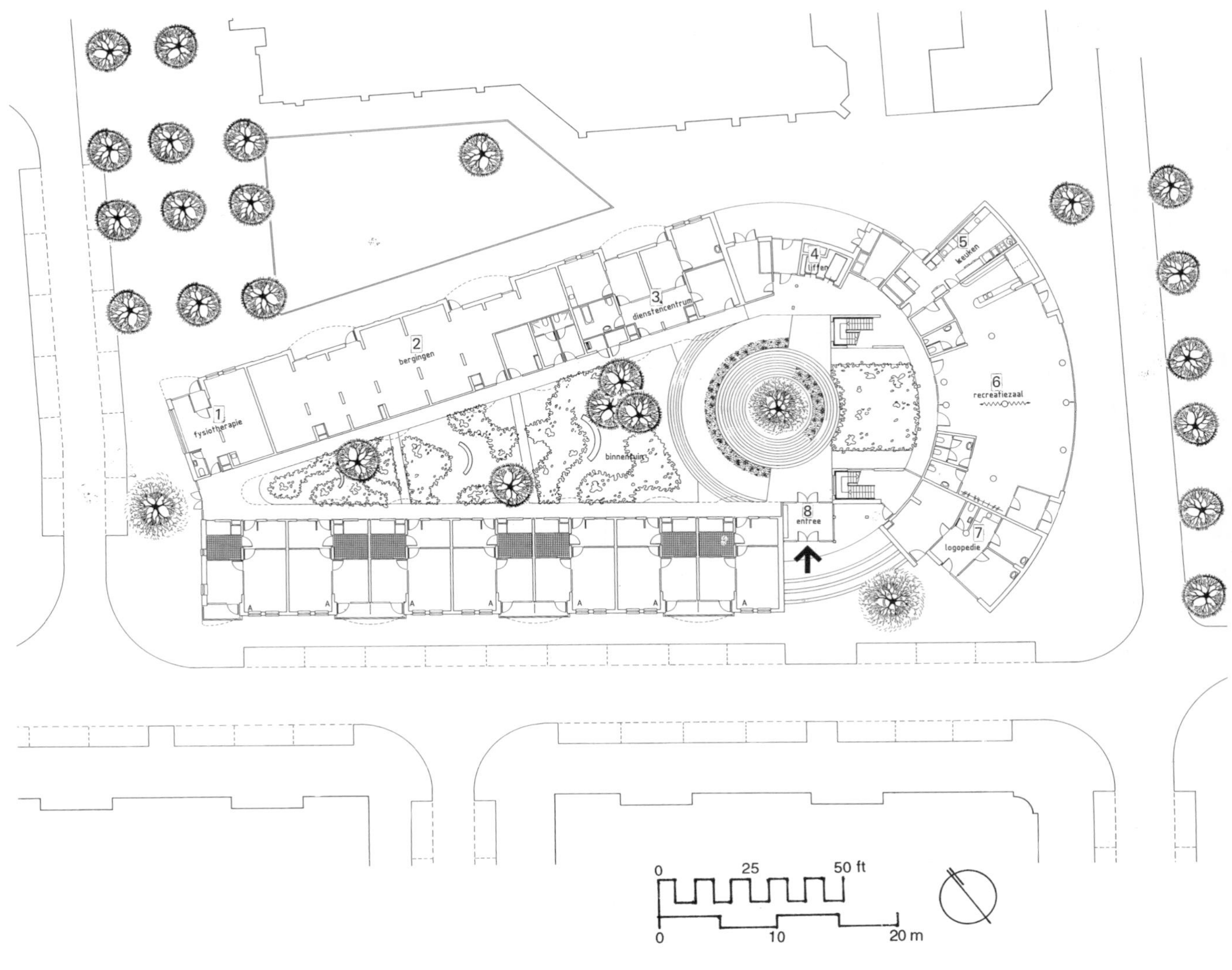

Key
1. Physical therapy
2. Storage
3. Administration offices
4. Elevator
5. Kitchen
6. multi-purpose room
7. Foot care
8. Entry

CASE STUDY DIAGRAM 5.3 *Ground Floor Plan*: Seven housing units located on the south edge face the street. Recreational and social activities for residents and neighbors take place in the multipurpose room at the building's east end. A circular paved patio with tables and seating is located near the entry to the atrium.

4. Unit Features

Because of the atrium, dwelling units receive light from two sides. A large three- by four-foot kitchen window faces the atrium, bringing light into the unit. Exterior walls have floor-to-ceiling glass that allows ample daylight into the bedroom and living room. A relatively large glass window connects the bedroom with the living room, allowing light to flow between the two rooms. The Dutch commonly use clear glass transoms above interior doors to allow borrowed light to reach interior rooms. No closets are planned in these units. The Dutch, like other Europeans, prefer armoires for storing clothing and personal items. The bathroom is sized and detailed to accommodate aging, disability, and handicap access. Emergency call buttons in the unit let residents call for help when needed.

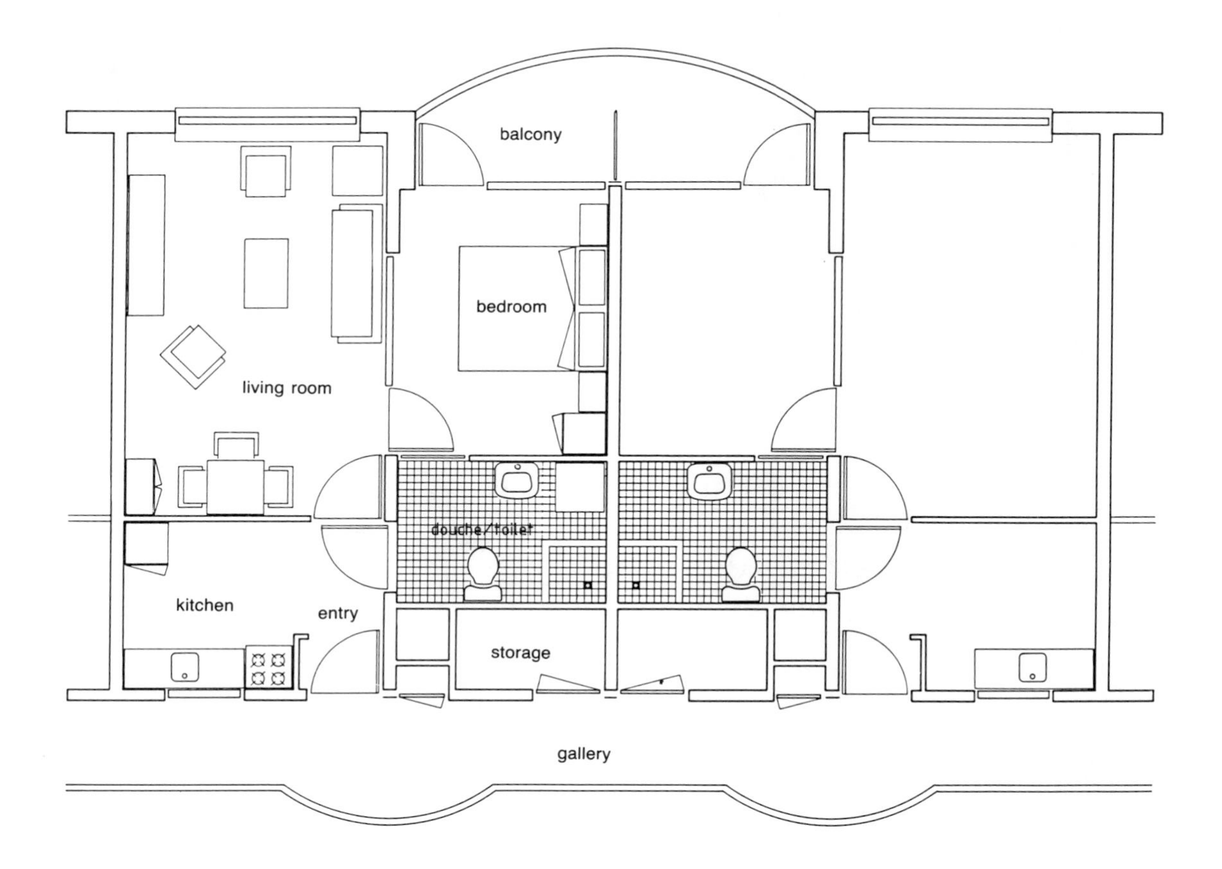

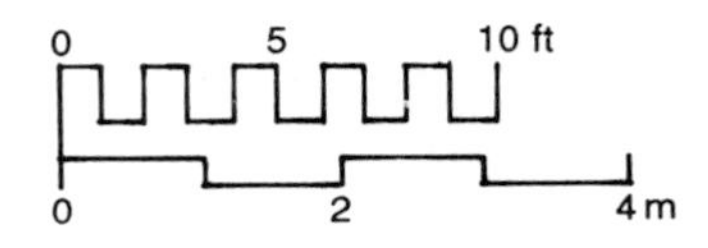

CASE STUDY DIAGRAM 5.4 *Typical Units*: Curved portions of the balcony corridor can accommodate a small table and chair. A glass window between the bedroom and living room visually connects these two rooms, giving the unit a spacious feeling. Curtains are used to assure privacy when desired.

5. Community Input During Design Process

The Dutch have an active community design process that solicits the input of neighbors about project attributes. Nonprofit housing development corporations are common and advocate this form of neighborhood communication. In the Jan van der Ploeg project, community residents helped to develop the architectural program were influential in advocating that the project benefit older people living in the surrounding neighborhood as well as project residents.

6. Interdependence

The open atrium and volunteer activities encourage interdependent helping behaviors. Approximately 30 percent of the residents take no services, while 15 percent are frail and require moderate levels of personal care assistance. Family members are also actively involved in helping residents with chores.

ROSEWOOD ESTATE, Roseville, Minnesota

2

Architect: Arvid Elness Architects, Inc.
Minneapolis, MN

Sponsor: Liberty Rosewood Limited Partnership
Minneapolis, MN

Building Characteristics:

a.	# of units	68
b.	# of stories	2 and 3
c.	Context	suburban
d.	Housing type	apt w/services
e.	Building parti/shape	rectangular
f.	Unit mixture	20—studio 48—1-bdr
g.	Size of most common unit (average)	550 sq ft
h.	Community facilities	no
i.	Community-accessible restaurant	no
j.	Year opened	1989

Residents Characteristics:

a.	Average age	82.5 years
b.	# of people	69
c.	# couples/men/women	4 couples 12 single men 49 single women
d.	% Bathing help	50%
e.	% Toileting help	30%
f.	% Incontinent	20%
g.	% Wheelchair-bound	41%
h.	% Cognitively impaired	6%

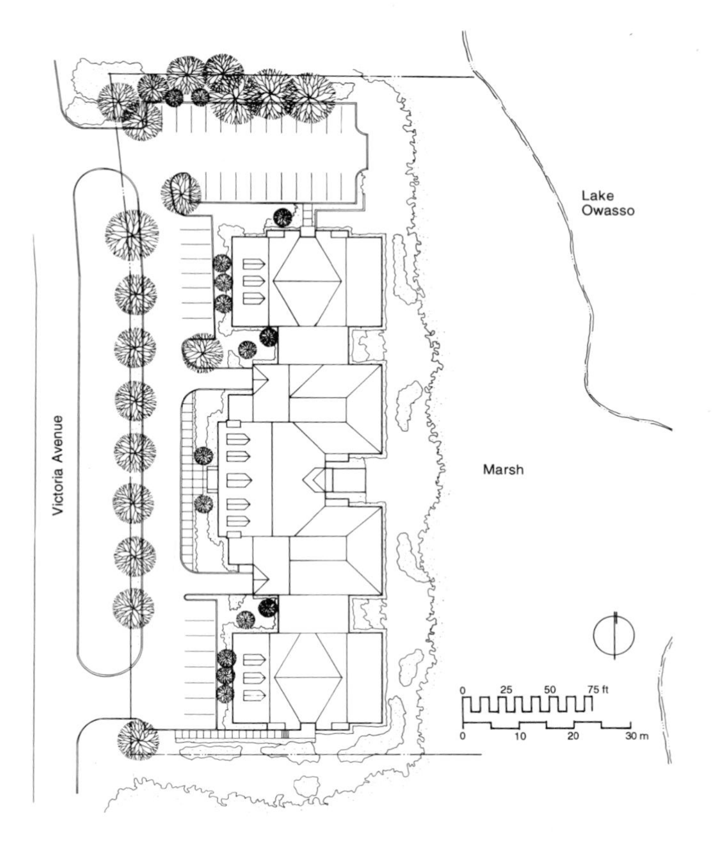

CASE STUDY DIAGRAM 5.5 *Site Plan*: *Rosewood Estate* appears in plan as three separate but connected buildings. Parking is located on the north edge of the site with guest parking occurring in front of the two end buildings.

Summary Description

Rosewood Estate is a sixty-eight-unit unlicensed alternative nursing home designed to operate on a home care delivery model. Nearly half of Rosewood's residents moved from a nursing home or hospital. Caregiving combines informal and formal services through a philosophy that emphasizes family relationships. The residential scale of the building combined with decentralized clusters of units reinforces a sense of interdependence and informal help between residents. A resident needs assessment, is carried out every six months by a caseworker, to establish a care plan that combines the help family members provide with formal service support. Personal care assistance is provided in fifteen-minute intervals for a cost of $5 an increment. Health care assistance is also time-shared. Residents can order help in advance or call for additional assistance any time it is needed through a Lifeline emergency call system. Units have full kitchens and bathrooms. Two thirds are one-bedroom apartments. The building, designed like a large house, has sloped roofs, colonial detailing and a three-part massing scheme. This gives it the appearance of a large, stately mansion home. The site overlooks a lake view on one side and an active highway on the other, giving residents a choice of views.

Project Features

1. Residential Appearance

Designed and constructed as a residential building, Rosewood appears very noninstitutional from the street. It utilizes a modest colonial style with simple detailing and a distinctive entry porch. The three-story building is nestled into a sloping hillside, two stories are visible from the upper street side and three stories at the rear of the site. Decorative dormers placed on the sloped roof alter the scale of the building and lessen its perceived height from the street. The building is divided into three small, simple masses connected in a linear sequence. Garage doors cover service areas where vans are stored, and contribute to the building's appearance as a residential dwelling. Lake views on one side and a relatively busy street on the other side give residents a choice of units with a passive or an active view.

FIGURE 5.3 *Rosewood Estate,* in Roseville, Minnesota, appears from the street to be a rambling colonial mansion house. The dominant center building uses an attached entry porch to give it focus. Dormers were added to the roof to create a sense of forced perspective, making the building appear smaller.

FIGURE 5.4 The Rosewood Room is designed to resemble an old-fashioned parlor, common in Midwestern housing at the turn of the century. The furniture and fireplace create a formal, dignified feeling. An adjacent three-season porch is far more informal in character and smaller in size, making it appropriate for intimate family conversations.

2. Family Involvement

The process of service assessment and care delivery at Rosewood involves a partnership between the family and the formal care provider. The family, in consultation with a case manager, can specify a blend of informal and formal services. Residents are interviewed to ascertain the type and level of assistance each needs, and are encouraged to do as much as they can for themselves. Formal personal care assistance is purchased from the home agency in fifteen-minute intervals for tasks that family members are not available to perform. A Lifeline emergency call system connected by mobile phone to attendants is used to augment the daily care plan by adding or reducing services. Because each fifteen-minute service increment is charged on a per task basis, residents and family members can save money by doing as much as possible themselves. The most common aid families provide includes help with laundry, medication organization, trips out of the facility, and walking assistance to the dining room for meals with family members. Residents can also save money by helping one another, which encourages interdependence among residents. Family events sponsored by Rosewood also increase family solidarity and enhance interfamily helping networks.

3. Interior Design Treatments and Features

Design features like fireplaces make reference to residential environments. Corridors that link the two contiguous wings are designed with offsets to avoid the perception of a long, undifferentiated hallway. An open stair near the entry allows residents to easily walk from one floor to another. The administration reception space, located on one side of the entry hall, uses an attractive, low-key approachable desk to make the first contact with visitors friendly. An unobtrusive corridor to one side leads to a suite of administrative offices. Several steps were taken to increase the amount of natural light that penetrates the interior, including eight-foot-high windows on the north facing the first floor and a two-story light well that borrows daylight from first-floor windows, allowing it to reach ground floor lounge spaces.

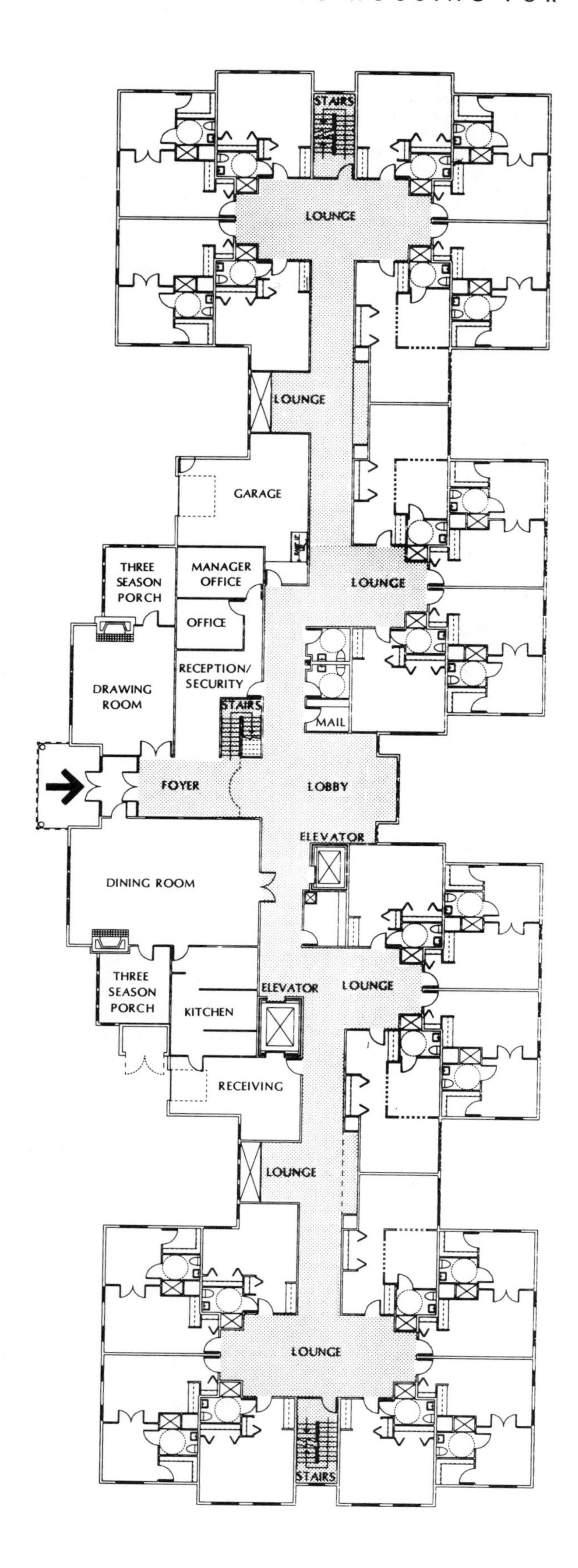

4. Flexible Service Delivery

Rosewood Estate is designed to keep residents healthy, independent, and out of a skilled nursing home for as long as possible. Hospice services can also be arranged with the participation of family members. Two meals are provided in a group dining room, but meals can also be delivered to a resident's unit for an additional charge. Approximately 30 percent of residents require no services when they move in, managing independently with security, meals, and help from family members. Currently, half the residents request help with bathing and 70 percent help with medication reminders. Assessments scheduled every six months update each resident's care plan.

5. Precedents Informed the Concept

The Elder Homestead project, in Minnetonka, Minnesota, was an important precedent for Rosewood. The home-care-based assessment method tested at Elder Homestead, revealed that the twenty-nine units in that project were too few to support the intensive array of services necessary to keep older frail residents independent. The unit count for Rosewood was increased to sixty-eight to improve operating efficiency. Attendants carry mobile phones linked to each resident's personal alarm system. This allows for instantaneous adjustments to the care plan.

6. Common Rooms Range in Character from Formal to Informal

Approximately 50 percent of Rosewood's gross building area is devoted to common space, lounges, and corridors. A range of common rooms from the formal Rosewood Parlor to an informal greenhouse are available for resident use. The dining room and central living room are the most popular common spaces residents use.

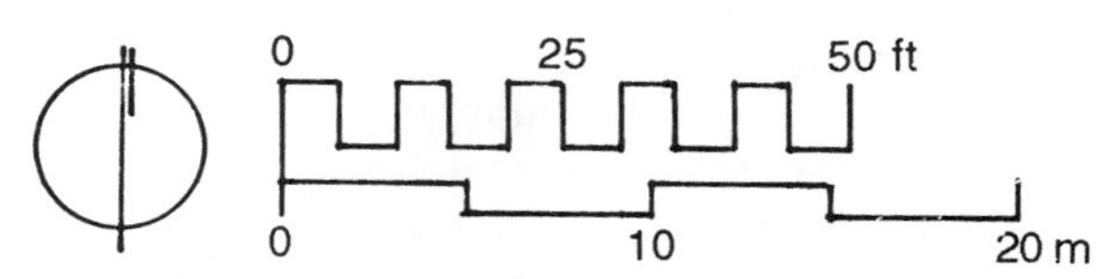

CASE STUDY DIAGRAM 5.6 *Entry Floor Plan*: Common spaces are clustered around the symmetrical entry. North and south buildings are treated as residential wings. Note how corridors have been offset and intermediary lounges located to avoid the perception of long corridors. The wide eight-foot corridors and the numerous undifferentiated lounges, however, make the building more anonymous and at times disorienting.

7. Residential Units

Dwelling units are designed to support an independent lifestyle. About two thirds are one-bedroom (550 square feet) and a third are studio alcoves (400 square feet). Each contains a complete kitchen and full bathroom. Bathrooms are designed with wheelchair-accessible showers and enough room for attendants to assist in transfers. A walk-in closet provides additional storage space. French doors with transparent glass panes connect the living room and bedroom, sharing natural light and increasing the feeling of spaciousness. Units are large enough to accommodate another person for an overnight stay.

8. Unit Clusters

Eight to ten units are clustered around decentralized lounges. The lounges provide shared social space and lend spatial variety to the corridors. However, these lounges often appear identical in style and configuration, making them monotonous and somewhat disorienting.

9. Problems to Be Resolved

Rosewood has pursued a number of intriguing ideas but has experienced some difficulties. Lacking a strong ground plane connection minimizes the role of landscape architecture in the daily life of residents. Interior design furnishings and details appear a little monotonous and the number and size of decentralized lounges seem greater than necessary. Additionally, the lack of a porte cochere at the entry makes it difficult to quickly drop off and pick up residents during inclement weather.

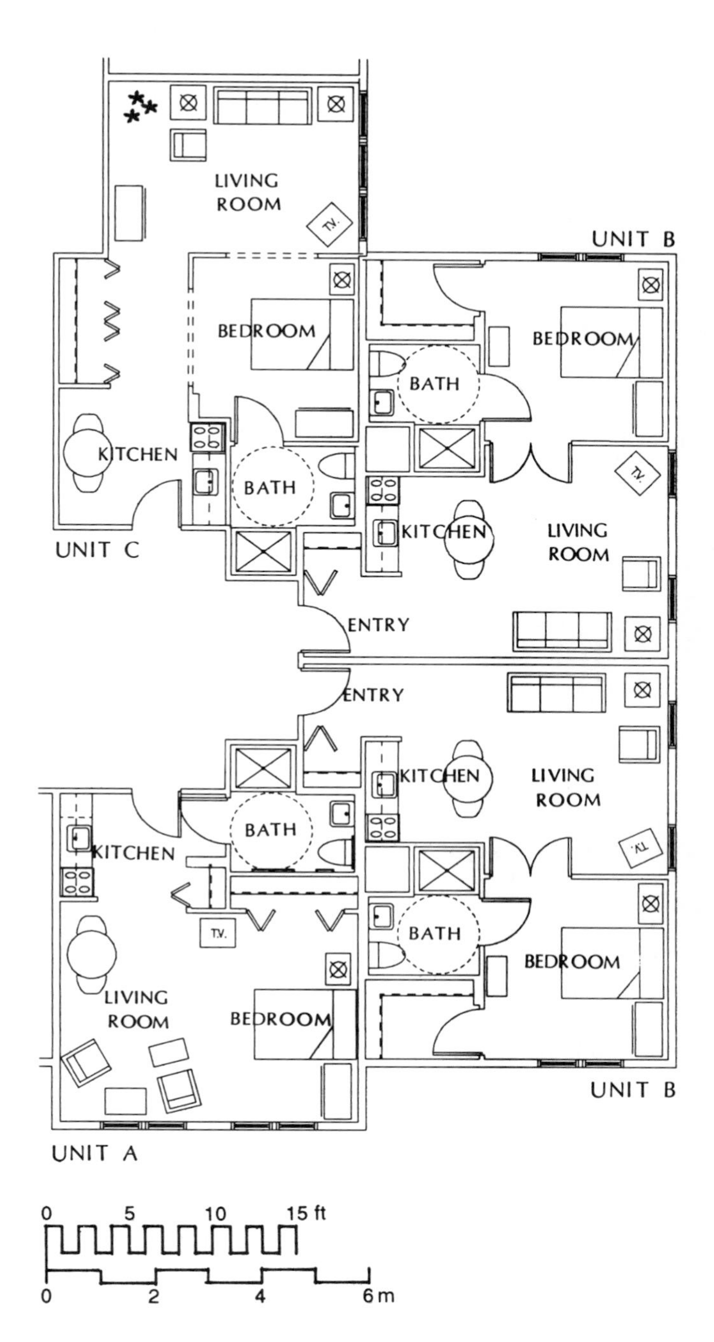

CASE STUDY DIAGRAM 5.7 *Unit Clusters*: Four units are typically clustered around an internal lounge. Careful planning has allowed these units to be fitted together with bedroom and living room spaces receiving perimeter lighting. The C unit uses a large cased portal opening from the bedroom to the living room to create greater identity for the bedroom in this studio unit.

BERGZICHT, Breda, the Netherlands

3

Architect: Architektenburo ir Frits Haverman bv
Breda, the Netherlands

Sponsor: Woningbouwvereniging Volkshuisvesting Breda
Breda, the Netherlands

Building Characteristics:

a. # of units	58
b. # of stories	4
c. Context	suburban
d. Housing type	steunpunt housing
e. Building parti/shape	enclosed street
f. Unit mixture	32 1-bdr 26 2-bdr
g. Size of most common unit (average)	650 sq ft
h. Community facilities	multipurpose room meeting rooms
i. Community-accessible restaurant	no
j. Year opened	1990

Residents Characteristics:

a. Average age	70 years
b. # of people	78
c. # couples/men/women	20 couples 10 single men 28 single women
d. % Bathing help	NA
e. % Toileting help	NA
f. % Incontinent	NA
g. % Wheelchair-bound	8%
h. % Cognitively impaired	3%

FIGURE 5.5 The three-story enclosed street provides a three-dimensional network of bridges, balconies, and corridors that facilitate the face-to-face meeting of residents and visitors. Walking down the corridor-street is a friendly and engaging experience. (Source: Sybolt Voeten)

Summary Description

This project, like Jan van der Ploeg, is also a steunpunt, or "point of support," housing arrangement, designed to provide limited nursing and home help service to residents as they age in place. The building contains an office for the district nurse as well as social and recreational spaces for residents and older people living in the surrounding community. Its most compelling design feature is an enclosed four-story gallery atrium designed to function as an enclosed street. The center space is covered by transparent, semicircular domes with temperature-controlled vents that open automatically to reduce overheating problems. Dwelling units have doors and kitchen windows that open onto the enclosed street. About half the units have a protected entry patio within the atrium, where plants can be tended. A network of balcony corridors provides access to units, and lets residents converse with one another between floors. The width of the atrium between the two housing blocks increases from 10.5 feet on the second floor to 11.5 and 12.5 feet on the third and fourth floors, allowing additional sunlight to reach the ground floor.

Project Features

1. Enclosed Street

The enclosed atrium street is a powerful concept that results in a friendly, socially conducive environment. Dwelling units have doors and windows that open onto this protected light-filled passageway, facilitating casual social encounters and stimulating the formation of friendships. Paired units on the west side of upper floors share an entry bridge and adjacent entry patios that further encourage informal socializing between neighbors. The network of bridges on upper floors are within comfortable conversation distance between floors, adding to the liveliness of the street. Upper floors receive plenty of light and are delightful. The ground floor, which is darker and contains only six units on the east side of the corridor, appears to be less successful. The west side of the ground floor contains the community room, which does not take full advantage of the enclosed street. The ground floor uses potted plants and a concrete tile floor to successfully give it the character and feeling of an outdoor street rather than of an interior corridor.

FIGURE 5.6 Small entry balconies to housing units on one side of the street create another way to activate the space. Residents water plants and sit outside greeting people.

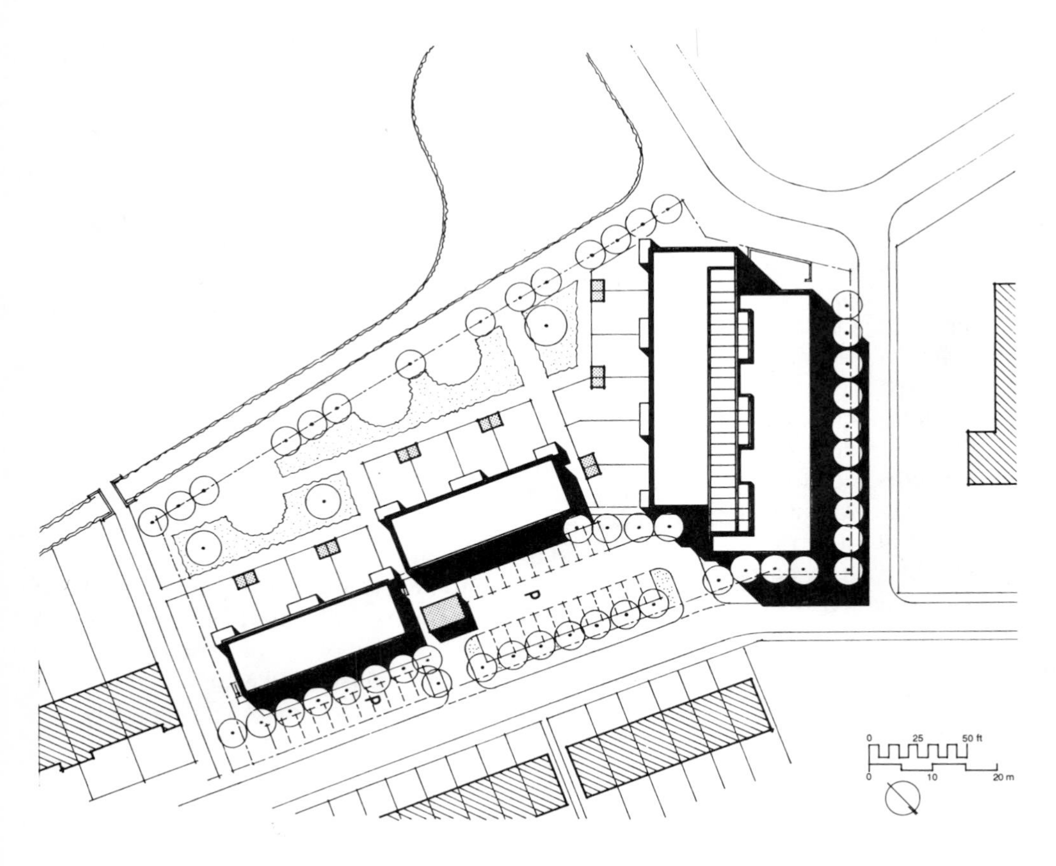

CASE STUDY DIAGRAM 5.8
Site Plan: The site is on the south edge of a small shopping district that serves suburban housing on the outskirts of Breda. Sixteen units of "lean-to" housing are located east of the building adjacent to a picturesque lake on the south edge of the site.

2. Indoor-Outdoor Spaces

Bergzicht offers three types of outdoor spaces. The first is the entry patio situated between the atrium and upper floor units on the west side of the project. This space, protected by the atrium, creates a greenhouse-like environment, which encourages plant growth. On the first floor, units on the east edge have large, generous patio gardens that are used for outdoor activities. Finally, all upper-floor units have small balconies accessible from the living room.

3. Location

Bergzicht is next to a market in a new suburban development to make it easier for residents to shop for food. The four-story configuration is an appropriate density considering its location next to commercial uses. The mixing of residents with neighbors in the community room gives this project the feeling that it belongs to the community instead of appearing to be private housing.

4. "Lean-to" Housing

This scheme contains sixteen units of *aanluen woning*—literally "housing that leans"—connected to the atrium building through a second-floor bridge. These units are similar in plan to the atrium units. Because it is two stories in height, a higher proportion of units have a garden. In other Dutch projects, lean-to housing has emergency call hardware linked to the central building and is situated near a service house, to facilitate the delivery of home care.

5. Temperature Control and Ventilation

The temperature of the enclosed street can be controlled through openable clear plastic domes that avoid overheating. Thermostatically controlled motors open the domes when temperatures reach a certain level or when smoke from a fire fills the space. Placing the pedestrian bridge on one side and making the width of the atrium larger at the top increases the amount of sunlight that reaches the lower floor.

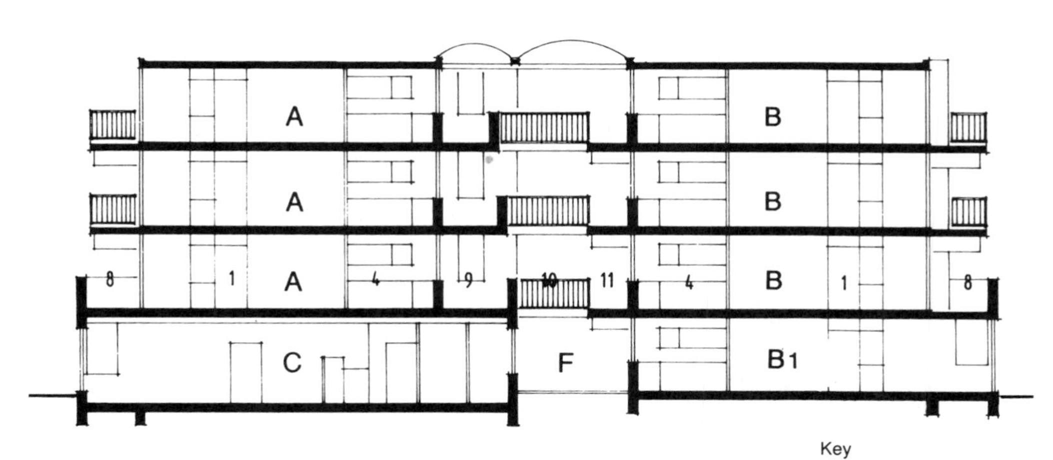

CASE STUDY DIAGRAM 5.9 *Building Section*: The enclosed street increases in width as the building increases in height. The north side of the first floor is devoted to recreational spaces and is open to neighbors as well as housing residents.

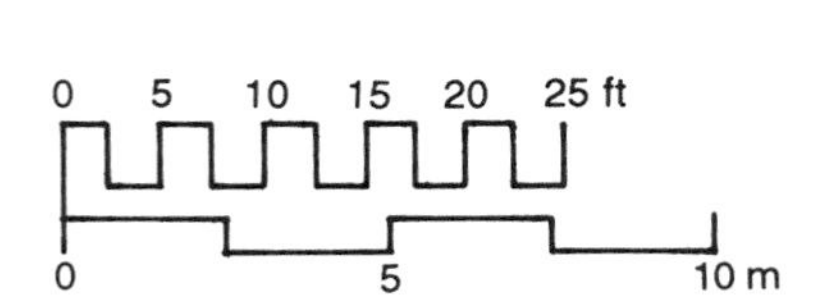

Key
A = Two Bedroom Units
B = One Bedroom Unit
B1 = One Bedroom Ground Floor Units
C = Community Spaces
F = Enclosed Street
1 = Living Room
2 = Master Bedroom
3 = Second Bedroom
4 = Kitchen
5 = Entry Foyer
6 = Bathroom
7 = Storage
8 = Balcony
9 = Entry Balcony
10 = Atrium
11 = Balcony Corridor

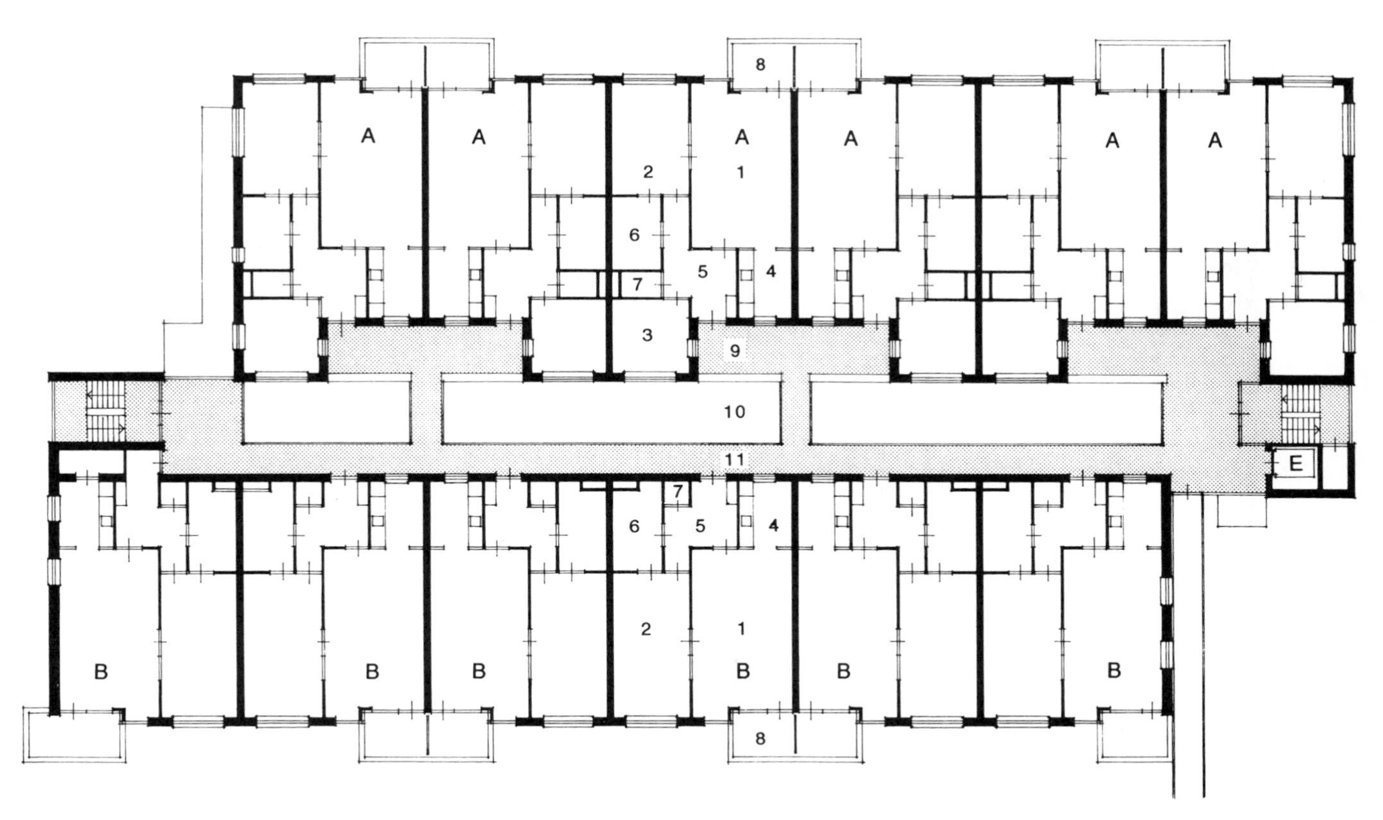

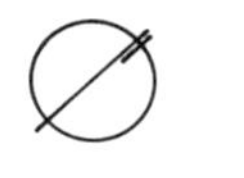

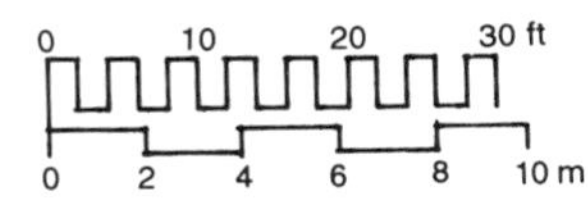

CASE STUDY DIAGRAM 5.10 *Second Floor Plan*: On each typical floor the enclosed street is flanked on one side by six two-bedroom units (A) and on the other by six one-bedroom units (B).

Key
A = Two Bedroom Units
B = One Bedroom Unit
B1 = One Bedroom Ground Floor Units
C = Community Spaces
F = Enclosed Street
1 = Living Room
2 = Master Bedroom
3 = Second Bedroom
4 = Kitchen
5 = Entry Foyer
6 = Bathroom
7 = Storage
8 = Balcony
9 = Entry Balcony
10 = Atrium
11 = Balcony Corridor

6. Community-Oriented Common Spaces

The common first-floor spaces are shared with neighborhood residents who enter through a separate outside door. This entry sequence isolates the space from the building, treating it as "neutral" rather than belonging only to residents. Residents and neighbors visit daily and participate in a range of social and recreational activities. The office of the district nurse is here, from which helping services for residents and neighbors are coordinated. Many of the activity organizers are volunteers who live in the project or in the adjacent neighborhood.

7. Unit Features

The project consists of a mixture of one- and two-bedroom units. The smaller second bedroom in the larger unit is located on the atrium street side. A kitchen window also opens onto the atrium. Typically, Dutch housing contains an entry space, which has doors to all other connecting rooms, including the living room. Glass transoms over these doors allow borrowed light to enter this "connecting" room. The living room and bedroom wall share a door and a transparent window panel, each of which is about three feet wide. The glass connection between bedroom and living room makes the unit appear larger. Those who prefer greater privacy in the bedroom can cover this window with a curtain.

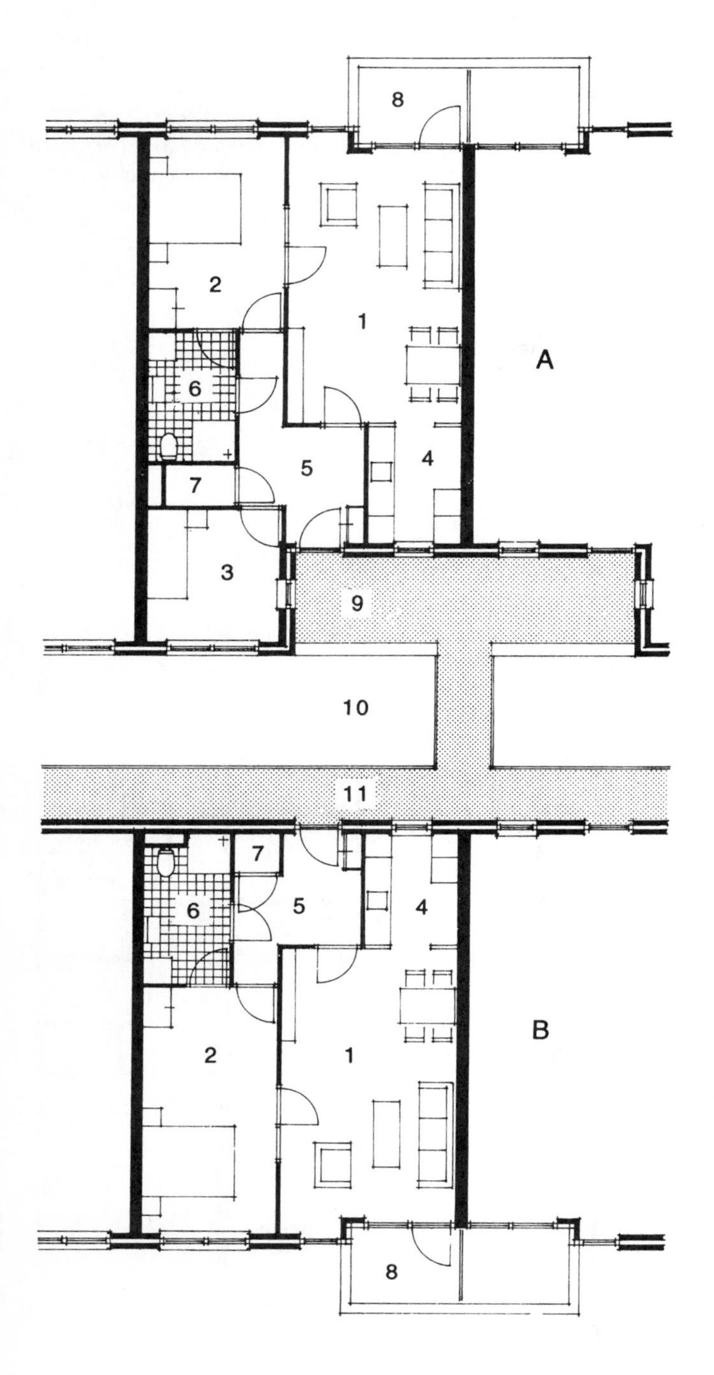

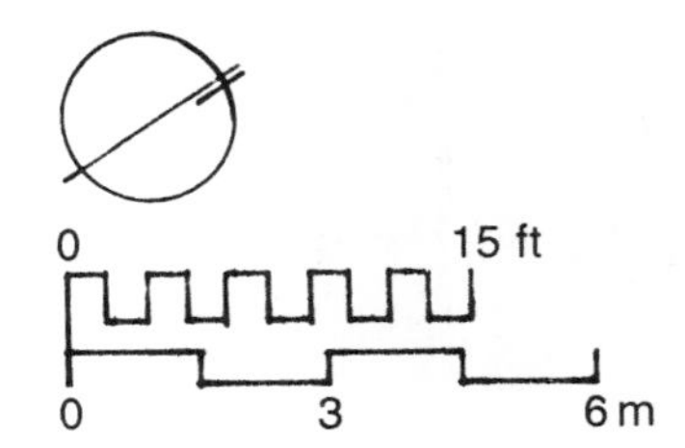

Key

A = Two Bedroom Units
B = One Bedroom Unit
B1 = One Bedroom Ground Floor Units
C = Community Spaces
F = Enclosed Street
1 = Living Room
2 = Master Bedroom
3 = Second Bedroom
4 = Kitchen
5 = Entry Foyer
6 = Bathroom
7 = Storage
8 = Balcony
9 = Entry Balcony
10 = Atrium
11 = Balcony Corridor

CASE STUDY DIAGRAM 5.11 *Unit Plans:* A unit contains both an entry porch and balcony. The B units on the first floor have a large outdoor area which can be used for growing flowers and vegetables. Note how the second bedroom of the A unit overlooks the enclosed street.

KUUSELAN PALVELUKOTI, Tampere, Finland

4

Architect: KSOY Arkkitehtuuria
Tampere, Finland

Sponsor: Tampere Elderly Home Association,
Tampere, Finland

Building Characteristics:

a. # of units	43 8 dementia units
b. # of stories	4
c. Context	suburban
d. Housing type	service house
e. Building parti/shape	atrium
f. Unit mixture	43 1-bdr 8 group home units
g. Size of most common unit (average)	500 sq ft
h. Community facilities	pool sauna restaurant
i. Community-accessible restaurant	yes
j. Year opened	1990

Resident Characteristics:

a. Average age	76 years
b. # of people	57
c. # couples/men/women	6 couples 7 single men 38 single women
d. % Bathing help	10%
e. % Toileting help	14%
f. % Incontinent	5%
g. % Wheelchair-bound	5%
h. % Cognitively impaired	20%

FIGURE 5.7 The atrium garden of *Kuuselan Palvelkuloti* is surrounded by a balcony corridor. In specific areas the corridor forms small spaces, that are thrust into the atrium with views on all sides. These sitting areas add variety and provide additional places for small group conversations.

Summary Description

This four-story, fifty-one-unit service house in suburban Tampere, Finland, is designed around an air-conditioned and heated garden atrium. Dwelling units located around an open balcony corridor overlook lush plant materials on the lower floor. Small group sitting areas attached to balcony corridors are thrust toward the center of the atrium space. The service house contains a number of amenities, including a swimming pool, restaurant, and therapy space, which are shared between residents and surrounding community members. A meals-on-wheels program, district "home health agency," and emergency response center provide outreach services to neighborhood residents.

A most intriguing addition to this service house is an eight-unit group home for Alzheimer's residents located on the top floor at one end of the atrium. This group home allows residents who experience major memory impairments to stay within the familiar context of the service house rather than be moved. The location at the top of a hill overlooking a lake is dramatic and beautiful. The restaurant is at one end of the building to capitalize on this view. The entry sequence to the project involves walking by social spaces, which are comfortable "perches" that overlook the atrium and encourage residents to engage in informal conversations. A kiosk-like building located on this platform provides snacks, tobacco, and cards. It is designed to resemble the traditional dark red, tin-roofed ice-cream kiosk found in urban parks in Finland. This activity brings a touch of whimsy to the entry deck while providing a place to purchase snacks.

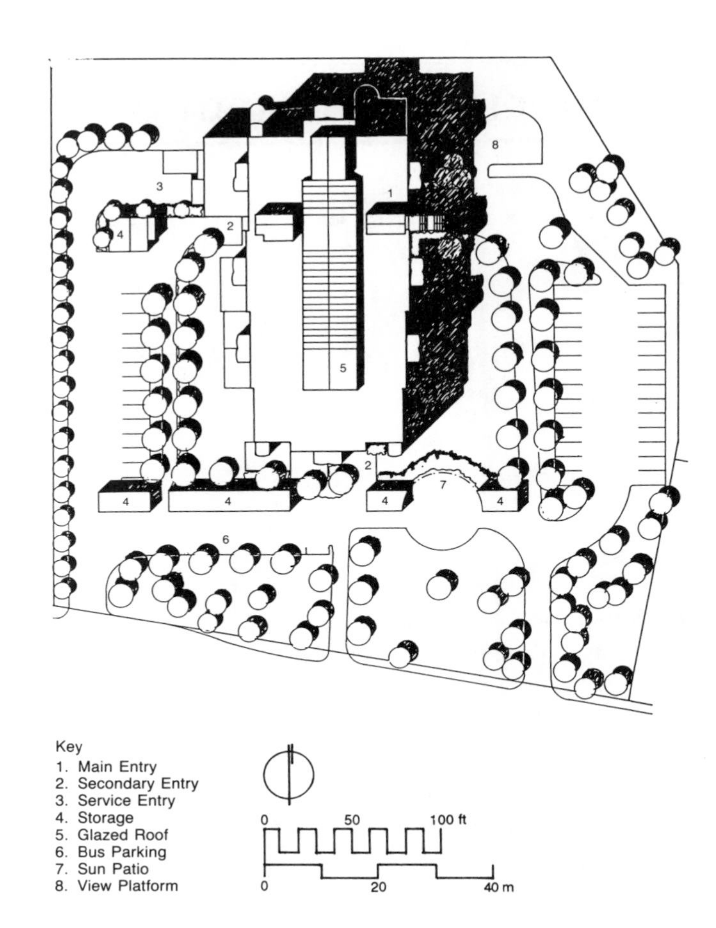

CASE STUDY DIAGRAM 5.12 *Site Plan*: The building is located on the northern edge of the site to facilitate a view of downtown Tampere and an adjacent lake. Adequate parking is provided on the east edge of the site for neighborhood residents who are also welcome here.

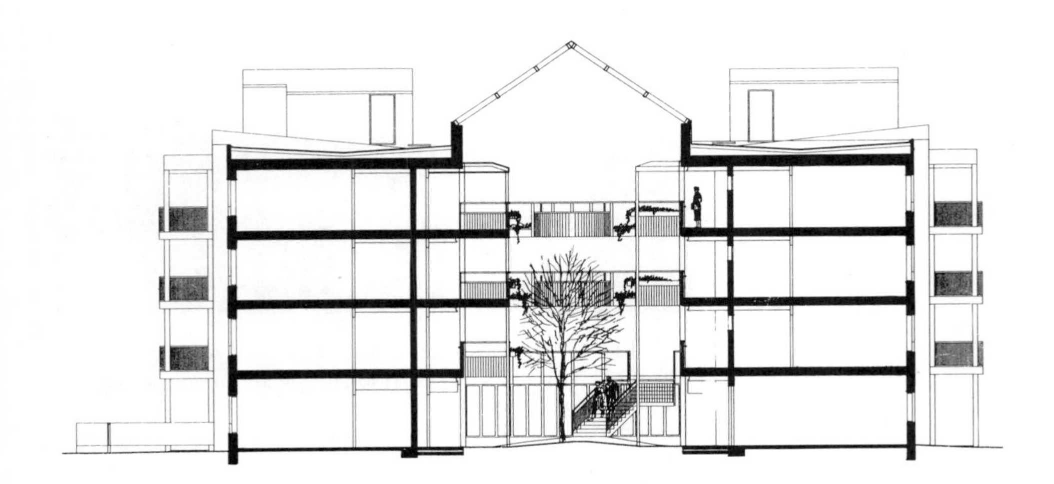

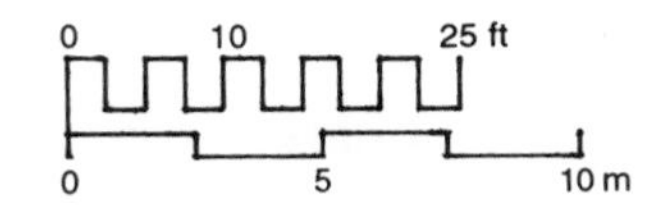

CASE STUDY DIAGRAM 5.13 *Building Section*: The garden atrium provides an attractive controlled view landscape for residents and visitors. Corridor balconies that provide access to units also open onto "perches" thrust into the air space above the atrium.

Project Features

1. Garden Atrium Creates Beautiful Controlled View

Units are clustered around a four-story central atrium. Two thirds of the ground floor is a garden with lush tropical landscape materials. The remaining third is an activity platform located above a swimming pool, both of which overlook the garden. Caged birds fill this landscaped garden with their melodious sounds. The glass atrium roof is designed to open when overheating becomes a problem; however, the atrium is also air-conditioned and heated. The quality of the filtered light that reaches the garden through the atrium is quite beautiful.

2. Site Location

The building is on a knoll above a lake, overlooking downtown Tampere, with a heavily wooded park in the foreground. The restaurant is designed with an outdoor patio that takes advantage of the view.

3. Two- to Four-Person Group Homes Are Located on the Top Floor

Two group homes that accommodate four persons each are located adjacent to one another on the south end of the top floor. Nurses and assistants available here on a twenty-four-hour schedule also respond to emergencies in the service house. The group home allows the service house to provide additional continuity of care. Residents who develop major mental impairments can stay within the familiar context of the service house. Thus, it can care for mentally and physically frail residents as they age in place.

4. Swimming Pool and Sauna Open to the Community

The Finnish people regard the sauna as a therapeutic and hygienic experience as well as a social and recreational one. It is a common feature in most homes for the aged. The swimming pool in this project is designed to be accessible to community members and relates to the atrium in a unique and clever way. Located adjacent to the atrium garden floor, light-level differences between the atrium and the enclosed pool during the day make the pool too dark to see into from the atrium. However, the sunlit garden view of the atrium from the pool is a transparent, tropical landscape. The window wall separating the pool from the atrium garden protects privacy while opening the pool to a beautiful controlled view.

FIGURE 5.8 The garden below has a rich collection of tropical and native plants and trees. Concrete sidewalks loop through the space adjacent to benches and tables in the garden. At one end is a large cage with songbirds that add sound and life to the place.

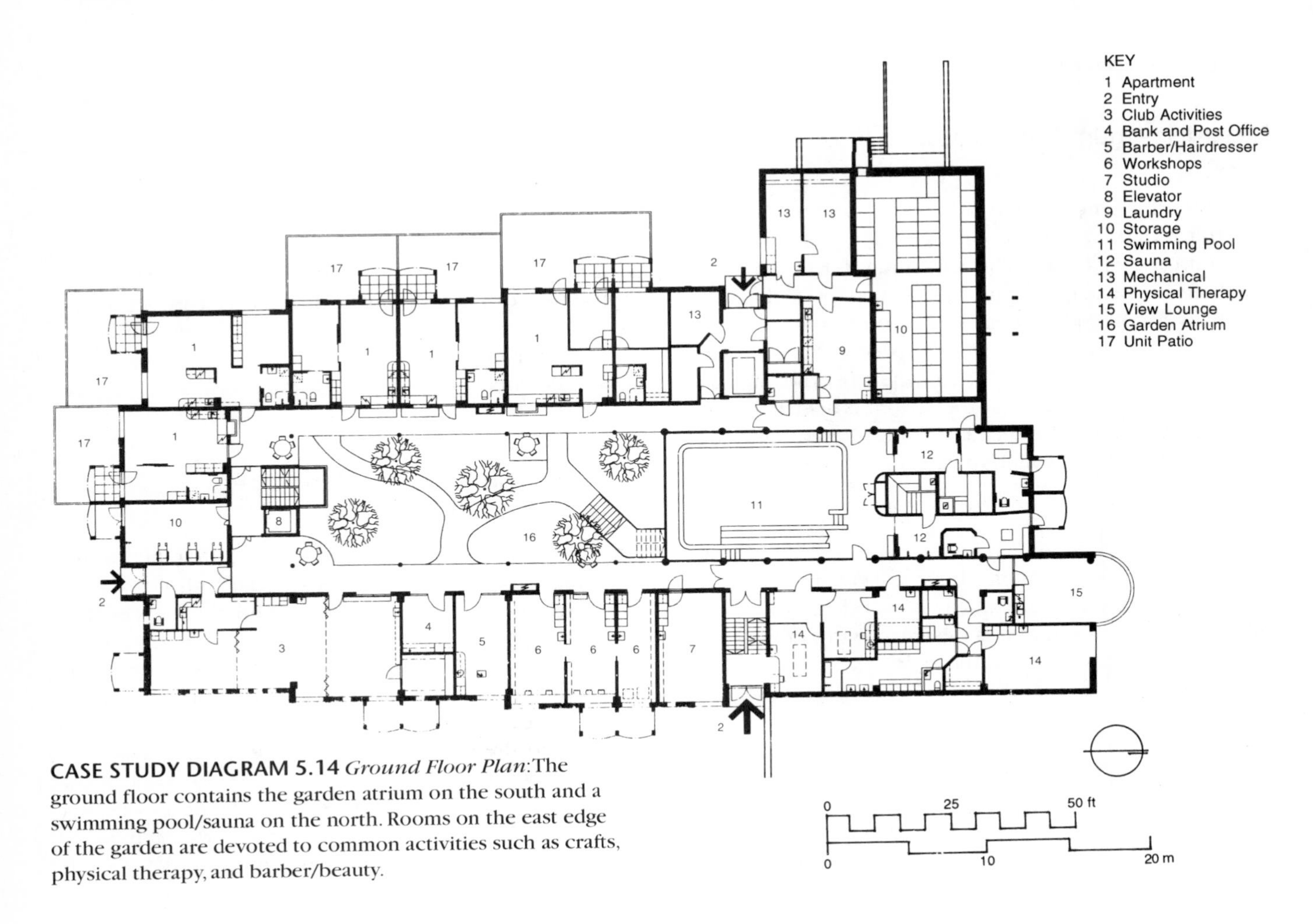

CASE STUDY DIAGRAM 5.14 *Ground Floor Plan*: The ground floor contains the garden atrium on the south and a swimming pool/sauna on the north. Rooms on the east edge of the garden are devoted to common activities such as crafts, physical therapy, and barber/beauty.

5. Snack Kiosk Effectively Uses Historic Reference

Upon entering the building you confront a freestanding kiosk at an oblique angle to the orthogonal grid of the building. This small structure with dark red vertical lap, wood siding, and a gray tin roof looks like an ice-cream kiosk. Located next to the front entry, the restaurant, and a popular sitting area, it provides residents with an excuse to visit these destinations. Residents are attracted here because the traffic generated by these destinations make the kiosk an attractive place to watch people and socialize.

6. Deck for Informal Socializing

Tables and chairs located adjacent to the entry door overlook the garden below. The seating is near the restaurant and the kiosk. This location allows residents to greet people entering and exiting the building and to engage in informal conversations in a relaxed way. Located in the center of the atrium, it receives a full complement of morning and afternoon sunlight. It is therefore an equally good place to pursue reading or watching activities.

7. Community Orientation

The building is designed to provide services to project residents as well as to older people living in the community. The restaurant and swimming pool are the most substantial shared amenities, but the lower floor contains occupational and physical therapy, barber and beauty shops, a crafts and ceramics studio, and a video room. District home help for surrounding neighbors is coordinated from the building and meals-on-wheels are prepared here and delivered to forty neighborhood residents.

8. Daylighting Is Very Effective

The low angle of the sun in Scandinavian winters makes daylight a scarce commodity. The atrium building maximizes daylight penetration into each dwelling unit. Windows from each unit borrow light from the atrium allowing daylight to enter each unit from two sides. Fire stairs also use transparent window walls, increasing daylight that enters the atrium and accommodating views to the outside.

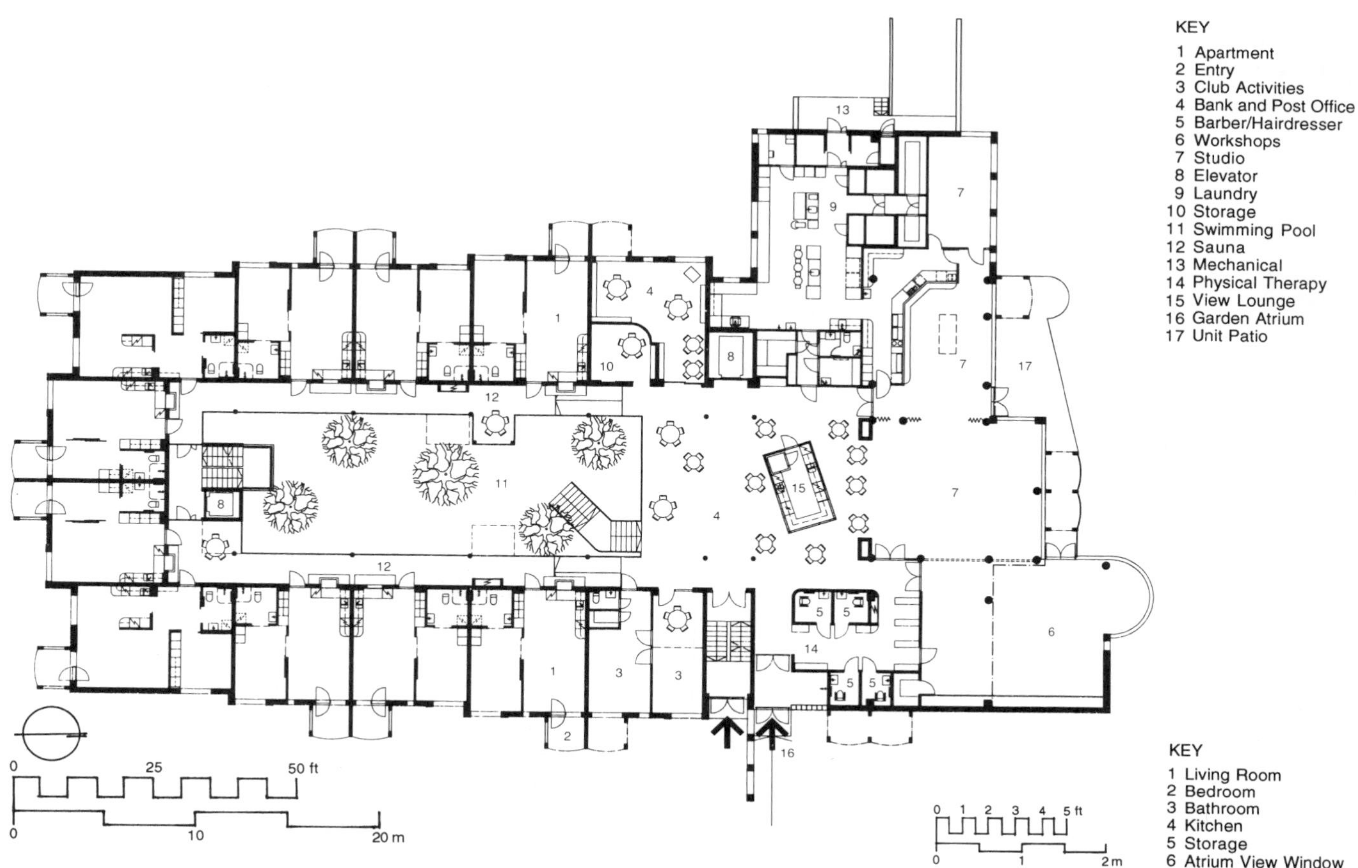

CASE STUDY DIAGRAM 5.15 *First-Floor Plan*: The main public entrance is on this floor. A meeting room and restaurant on the north end overlook the view. An entry deck located above the pool overlooks the garden atrium below. Note the "floating" kiosk building. Its stylistic treatment and skewed placement adds to the informal character of this busy space.

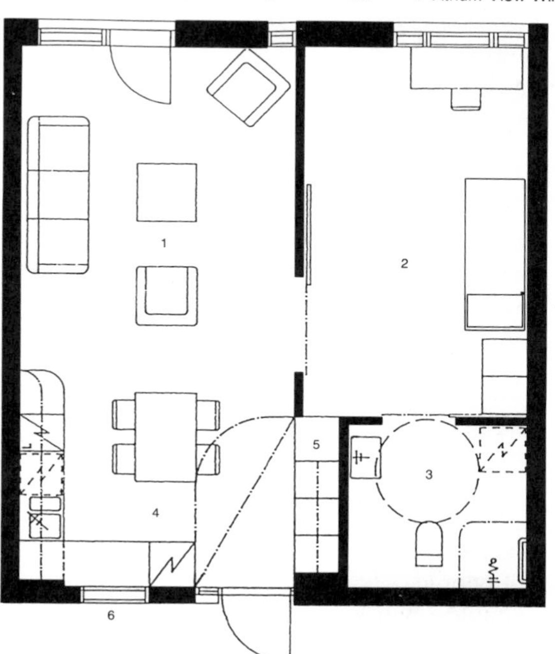

CASE STUDY DIAGRAM 5.16 *Units*: An L-shaped kitchen adds flexibility for furnishing the unit. Note the window that provides the kitchen with a view of the atrium. Sliding doors link the bedroom with the living room and bathroom. Note how built-in armoires rather than closets are used for storage.

9. Unit Features

In general, Finnish dwelling units measure 10–15 percent smaller in size than their Swedish and Danish counterparts. A small greenhouse window above the kitchen sink projects into the balcony corridor, providing a view of the atrium. Each unit has a door bell and mailbox to clarify its identity as a front door. Residents can personalize their unit by installing their own light fixtures, replacing floor and wall coverings, and bringing their own furnishings. These detailed personal changes are common in European rental housing because residents generally have longer housing tenure than their counterparts in the United States. Kitchens are L-shaped and designed to be adaptable. Toilet rooms have sliding doors that accommodate wheelchair access.

NYBODERGAÅRDEN, Copenhagen, Denmark

5

Architect: Ib and Jorgen Rasmussen Arkitekter
DSB Station, Denmark

Sponsor: Nybodergaården—An independent nursing home with an agreement from the municipality of Copenhagen

Building Characteristics:

a. # of units	54
b. # of stories	3 + Bsmt
c. Context	urban
d. Housing type	plegehem
e. Building parti/shape	atrium (activity)
f. Unit mixture	48 studios 6 residential units
g. Size of most common unit (average)	260 sq ft
h. Community facilities	physical therapy occupational therapy day care
i. Community-accessible restaurant	no
j. Year opened	1977

Resident Characteristics:

a. Average age	82 years
b. # of people	54
c. # couples/men/women	17 single men 37 single women
d. % Bathing help	83%
e. % Toileting help	83%
f. % Incontinent	50%
g. % Wheelchair-bound	31%
h. % Cognitively impaired	35%

FIGURE 5.9 The *Nybodergaården* first-floor atrium is subdivided into several different areas for eating and socializing. Near the entry is an informal socializing area. The object on the left side of this photograph is a bar where drinks and snacks are sold every afternoon. At the rear of the atrium is a dining room. Natural light penetrates the center of the building and single-loaded corridors on the upper floors allow residents to overlook the activity below.

Summary Description

Nybodergaården is a Danish plegehem, or skilled nursing home. It differs from the United States standard in several significant ways. Each of the 260-square-foot rooms is single-occupied, with a full bathroom and a small refrigerator. Physical and occupational therapy are stressed as an important component of the plegehem lifestyle. The emphasis on therapy is so pronounced, one has the impression the plegehem is part rehabilitation hospital. Paradoxically, an open bar on the first floor of the atrium, which sells alcoholic drinks and snacks to residents, neighbors, and families during the day, imbues the setting with a relaxed social atmosphere. Meal service is decentralized, with several smaller, more intimate places in which residents can eat.

Nybodergaården is located in the historic Nyboder district and its overall shape conforms to the long, narrow housing blocks constructed here in the late 1500s. In this building, the traditional open-air courtyard between the narrow housing blocks has been transformed into a covered light-filled atrium. The traditional ocher color of the building is consistent with neighboring dwellings, furthering its camouflaged fit into the surrounding context. The atrium design fills the center of the project with light, facilitating group activities on the first floor. Upper floors have wide balcony corridors that overlook the atrium. Residents have furnished these corridor spaces as extensions of their dwelling units with chairs, tables, and occasionally a beach umbrella to protect them from the sun.

Project Features

1. Atrium Design Daylights Interior Spaces

The transparent peaked roof over the center of the building allows natural daylight to reach activities on the ground floor. Corridor balconies on the sides overlook the atrium, creating a visual connection between resident rooms on the upper floor and ground floor activities. Some overheating problems exist in the summer but in general the scheme works well.

2. Building Fits the Context Well

In the historic Nyboder district of Copenhagen where Nybodergaården is located (an area developed in the late sixteenth century and planned as housing for the merchant marine), the old building form consists of two parallel, extruded two-story housing blocks separated by an outdoor garden. The Nybodergaården project continues this configuration by transforming the center garden into a linear atrium. From the outside it resembles the sixteenth-century housing block in form and character, utilizing the same yellow-ocher coloration of the old buildings. The adaptation of the old form allows this new building to fit unobtrusively into the context.

FIGURE 5.10 The surrounding neighborhood contains housing blocks in a pattern that has been maintained for hundreds of years. This project was newly constructed, but took its form, coloration and scale from the surrounding urban pattern. It is creatively transformed into a more functional housing arrangement for this particular user group.

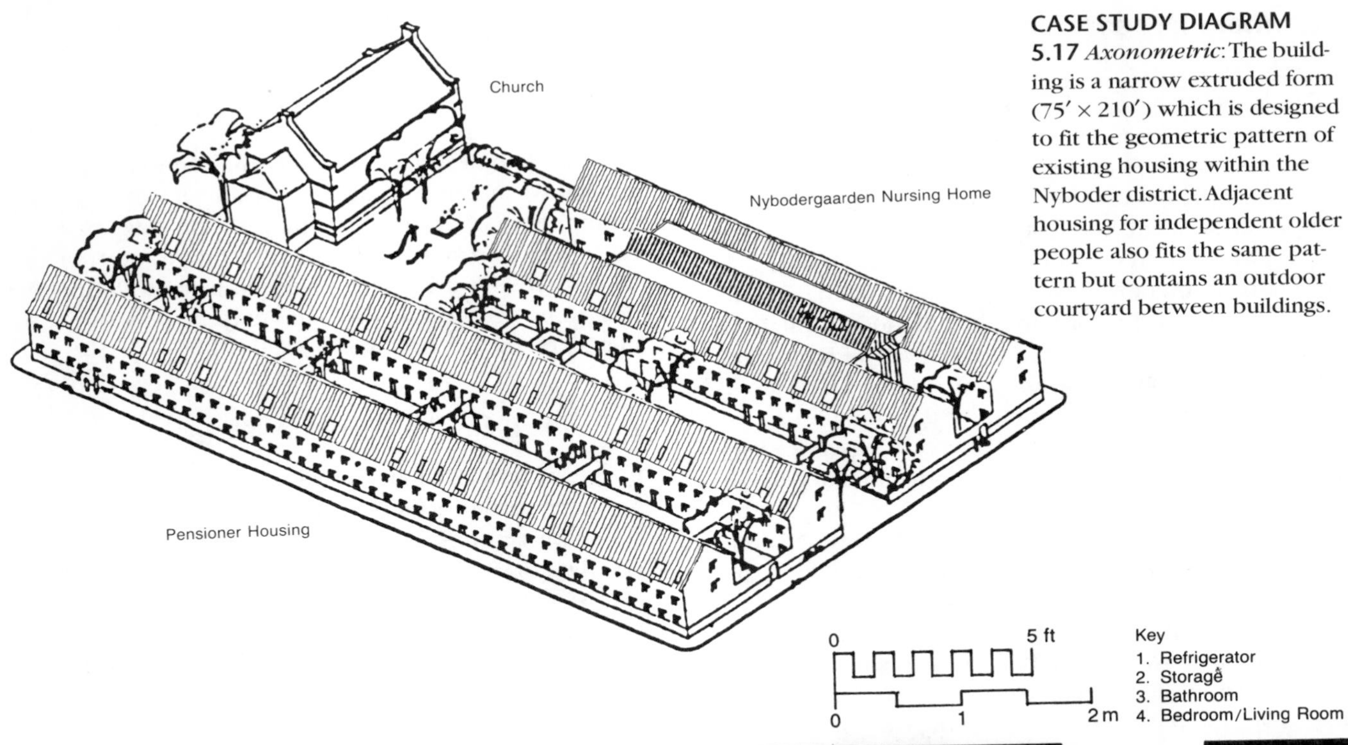

CASE STUDY DIAGRAM 5.17 *Axonometric*: The building is a narrow extruded form (75′ × 210′) which is designed to fit the geometric pattern of existing housing within the Nyboder district. Adjacent housing for independent older people also fits the same pattern but contains an outdoor courtyard between buildings.

3. Unit Features

Resident rooms are private, single-occupied, and contain a full bathroom. Although residents don't have food preparation equipment, each unit has a refrigerator and residents are welcome to bring their own furniture. A hospital-type bed is furnished by the facility to minimize staff back injuries. A hinged eight-inch side panel can be opened to facilitate moving a bed in or out of the room. Wood entry doors are thirty-six inches wide. An eight-foot-wide balcony corridor overlooks the atrium, providing a convenient and interesting "patio" extension of the resident's unit. Many residents have placed chairs, tables, and even umbrella tables by the rail overlooking the atrium.

4. Emphasis on Physical and Occupational Therapy

Physical and ergotherapy (occupational therapy) are an important daily activity. Residents are engaged in building and maintaining physical competency. Very few residents are bedfast, partly due to management policies that get residents out of bed and stress involvement in a range of activities every day. Therapy programs are also open to older people living in the surrounding neighborhood. Considerate design features used for walking therapy include stairs designed with low risers and wide treads (five-inch rise/thirteen-inch run) to make it easier for older residents to walk from floor to floor.

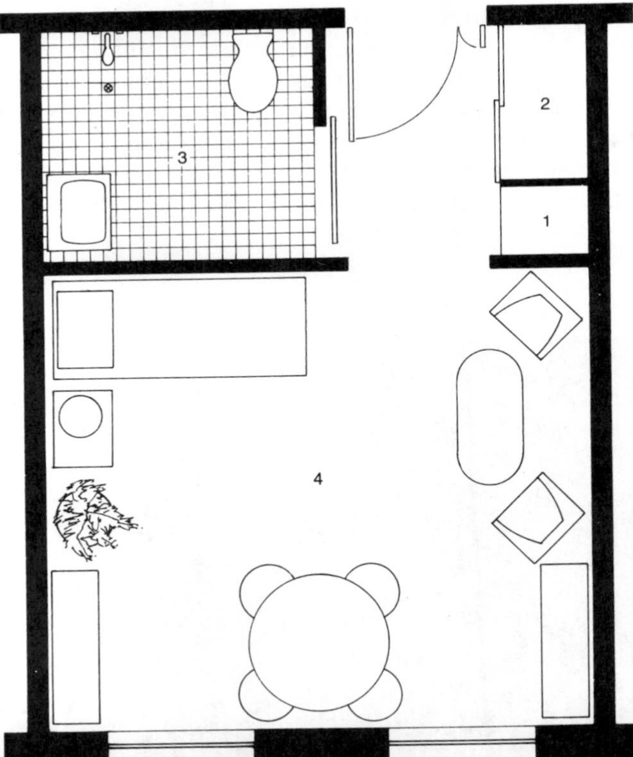

CASE STUDY DIAGRAM 5.18 *Unit*: Each unit has a small refrigerator for food. Large accessible bathrooms with sliding doors facilitate resident transfers. The unit entry door is designed in two pieces to accommodate the moving of a bed into or out of the room. Residents bring their own furniture.

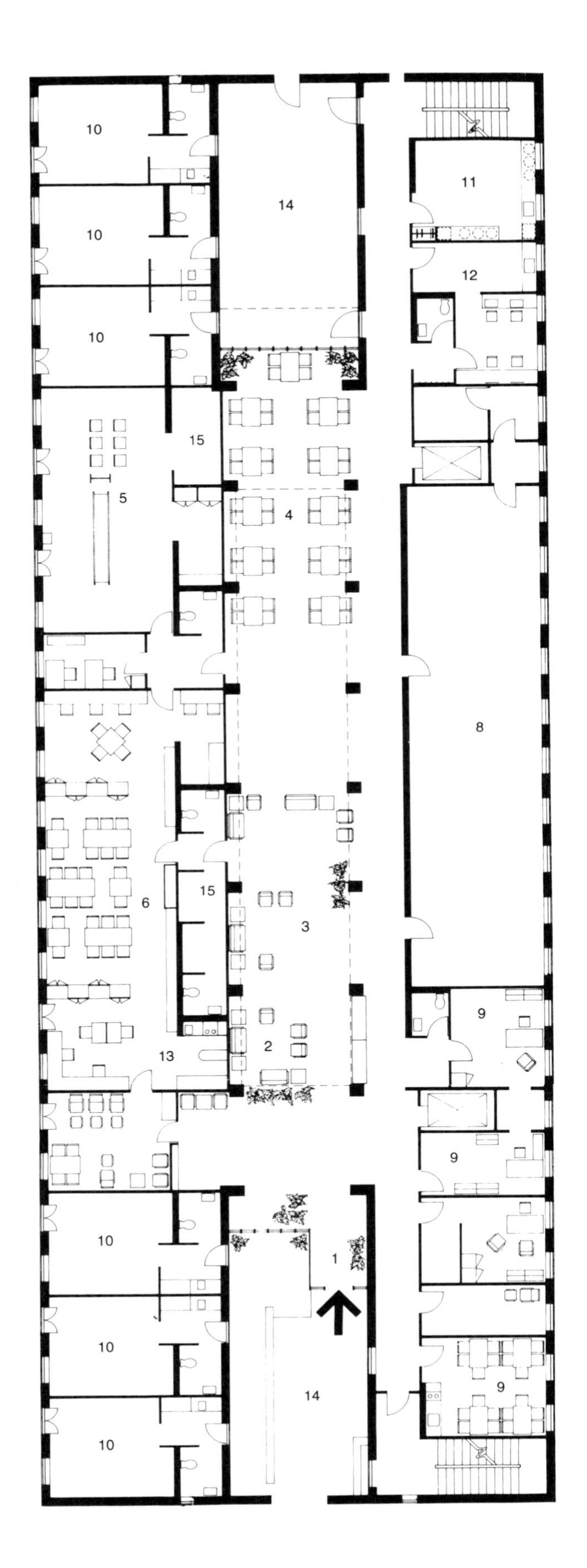

5. Programs for Neighborhood Residents

General housing quality in this part of Copenhagen is relatively low. A number of older people live in substandard stock with no bathing facilities. Nybodergaården provides day-care services and hygiene facilities for nine community residents. A training kitchen designed for day-care recipients is also shared by residents who participate in this form of therapy.

6. Bar on First-Floor Is Open to Residents, Neighbors, and Family Members

Drinks, including beer, wine, and liquor, as well as snacks are sold at an open bar located on the ground floor of the multistory atrium near the entry. The bar activates a lounge located here. Laughter, conversation, and the convivial behaviors present here form a radically different impression of long-term care in Denmark. This behavioral setting gives the first floor a relaxed, informal atmosphere that promotes social exchange.

7. Housing Mixed with Health Care

In addition to the forty-eight rooms for heavier care patients, Nybodergaården also offers six separate apartments for monitored care. These units have outdoor entries accessible from forecourts in the front and rear of the project. Each forecourt is connected to the street. Each flat has a small kitchen.

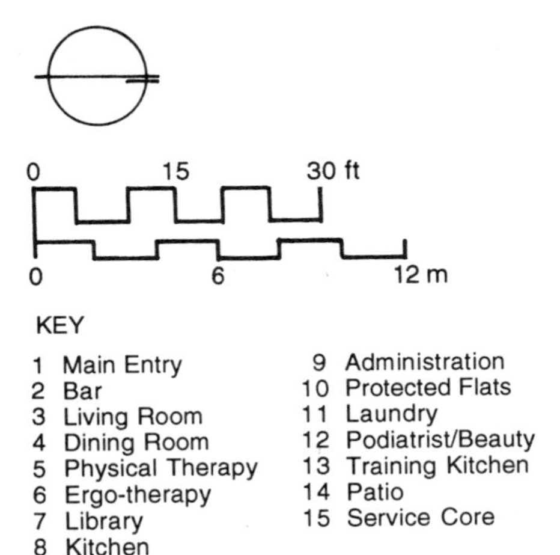

CASE STUDY DIAGRAM 5.19 *Ground Floor Plan*: The central atrium is used for dining, recreation, and social activities. The south edge of the ground floor is devoted to therapeutic activities. The east and west end of the building contain six apartments with small kitchens for independent residents who need some assistance. Entry courts in the front and back provide outdoor places to sit.

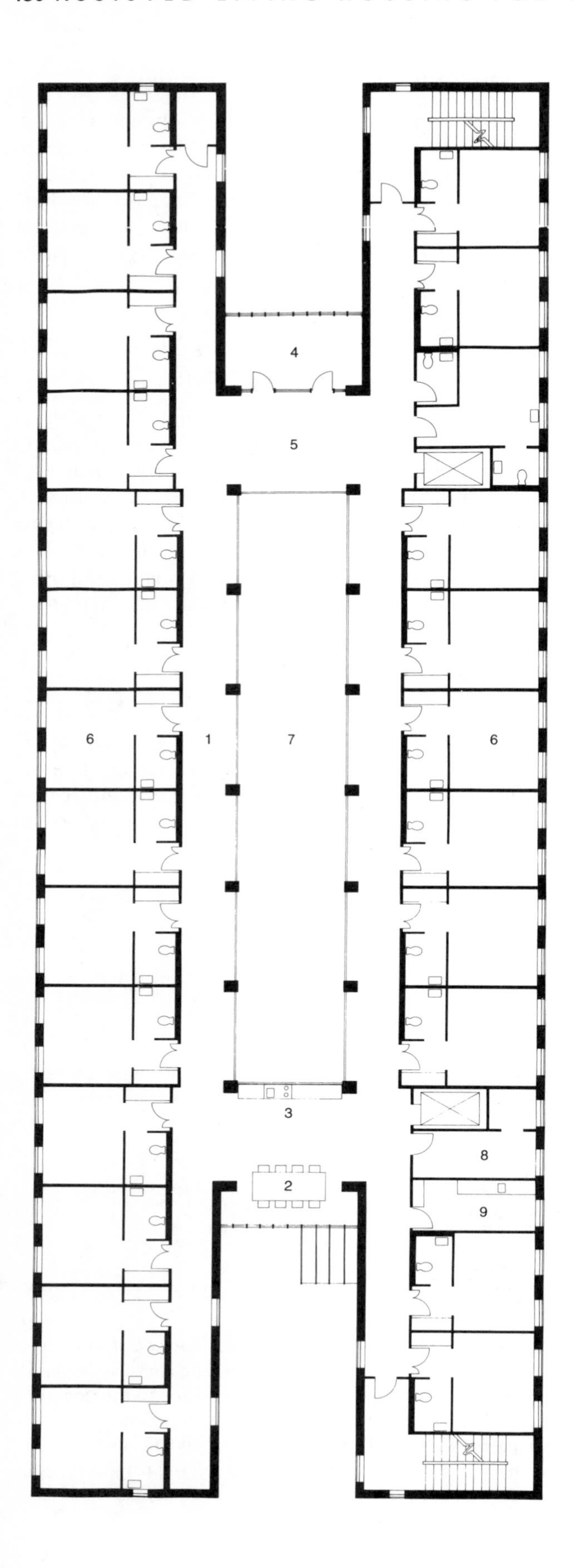

8. Nurses' Station

Overlooking an exterior courtyard with a greenhouse window and a table for staff meetings, the nurses' office is also unlike the typical American nurses' station. Nonthreatening and approachable, round tables outside the office provide a place for interviews and conversations with residents. In other Danish plegehems, the nurses' station is in a corridor alcove or is simply a large table placed in a wide portion of the corridor, where staff members can meet with residents to discuss problems. Instead of creating barriers between residents and staff, they have chosen locations and furniture arrangements that encourage interaction and contact.

9. Decentralized Service Care Organization

One of the compelling aspects of the plegehem is the effort made to care for project residents in small groups. With forty-eight nursing units, the scale of the overall project is intimate. However, this intimacy is further increased by grouping twenty-four units on two upper floors. These floors are further split into fourteen and ten resident working groups. Residents have three choices in taking meals. They can eat in their room, in the small café dining room on the ground floor, or around a small table at the east end of each resident floor.

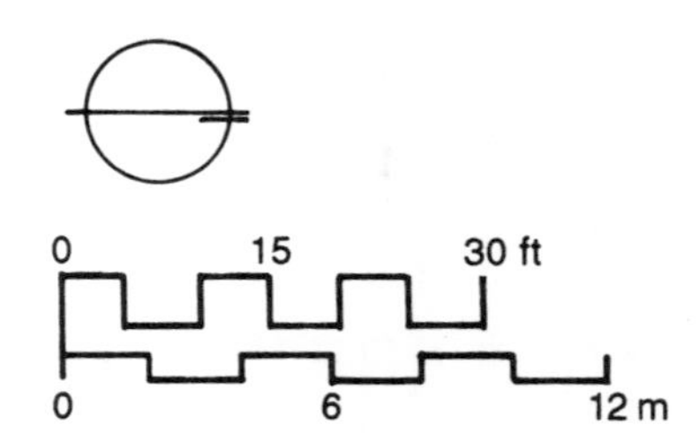

KEY

1 Balcony Corridor	6 Typical Unit
2 Dining Alcove	7 Open Atrium
3 Open Serving Area	8 Linen Storage
4 Nurses Station	9 Scullery
5 Social Lounge	

CASE STUDY DIAGRAM 5.20 *Typical Floor Plan*: Residents can take meals on the first floor, in their rooms, or at a small table on each floor that overlooks the entry. Balcony corridors adjacent to units allow each resident to have a place to view atrium activities below. A nurses' station at the end of each floor opens to a lounge where tables and chairs facilitate discussion.

SUNRISE RETIREMENT COMMUNITY, Fairfax, Virginia

Architect: Heffner Architects
Alexandria, VA

Sponsor: Sunrise Retirement Homes and Communities
Oakton, VA

Building Characteristics:

a. # of units	47
b. # of stories	3 + Bsmt
c. Context	suburban
d. Housing type	home for aged
e. Building parti/shape	L shape
f. Unit mixture	35 studio 12 1-bdr
g. Size of most common unit (average)	275 sq ft
h. Community facilities	no
i. Community-accessible restaurant	no
j. Year opened	1990

Resident Characteristics:

a. Average age	82 years
b. # of people	60
c. # couples/men/women	9 single men 51 single women
d. % Bathing help	100%
e. % Toileting help	25%
f. % Incontinent	25%
g. % Wheelchair-bound	10%
h. % Cognitively impaired	40%

FIGURE 5.11 The open stair is a *Sunrise* signature element. It enhances the entry experience while adding to the spatial hierarchy of the building and creating a memorable place for orientation. The stair is also used for exercise. A receptionist sits at the desk and greets residents and visitors.

Summary Description

This three-story, forty-seven-unit assisted living facility uses the imagery of a Victorian mansion house as the basis for its design concept. Sloped roofs, dormer windows, a low-slung attached porch, corner turrets, and a low-scaled porte cochere give it a friendly and approachable look from the street. Inside, residential references two-story entry contains a dramatic open stair that links the second-floor with the ground floor. The use of antique furniture and wall-mounted displays of everyday items from the childhood days of residents are memory-provoking for residents and fascinating to visitors. Art has been chosen that depicts emotionally positive themes such as intergenerational relationships. Art content adds an affective dimension to the environment.

Dwelling units and common spaces are designed in scale and function to resemble a bed-and-breakfast inn rather than an institution. The resulting scale of common rooms and corridors appears intimate and manageable. The L-shaped building configuration reduces perceived corridor length. A residentially scaled room at the end of each corridor serves as a semiprivate living room around which units are clustered. The care philosophy is oriented toward keeping residents independent and out of nursing homes for as long as possible. "Miniphysicals" are given weekly to track existing conditions and anticipate developing problems. Creative therapies involve residents with pets and children. Dwelling units are generally small and without food preparation equipment.

Project Features

1. Residential Appearance

The building is designed to resemble a three-story mansion. Porches, covered entries, and a dormer roof on the third floor reduce the perception of building mass and height from the street. Sloped roofs, residential materials, and friendly design details are employed to give the building a comfortable, homelike appearance. Octagonal-shaped corner turrets add exterior variety and uniqueness while creating sitting alcoves for several unit interiors. The L-shaped building configuration focuses attention on the entry and helps define the front lawn as a landscaped entry court.

0 5 10 15 ft
0 2 4 6 m

CASE STUDY DIAGRAM
5.21 *Third-Floor Plan*: The *Sunrise* plan minimizes long corridor lengths by providing a living room social space at the end of each corridor. The symmetrical, compact plan minimizes travel distance from the unit to the elevator. The resident kitchen is used to support group activities.

FIGURE 5.12 The residential scale of the *Sunrise* at Fairfax is reinforced through the use of dormer windows on the third floor, attached porch elements on the first floor, and residential detailing throughout the building. The mansion house imagery gives it the feeling of a house rather than a nursing facility.

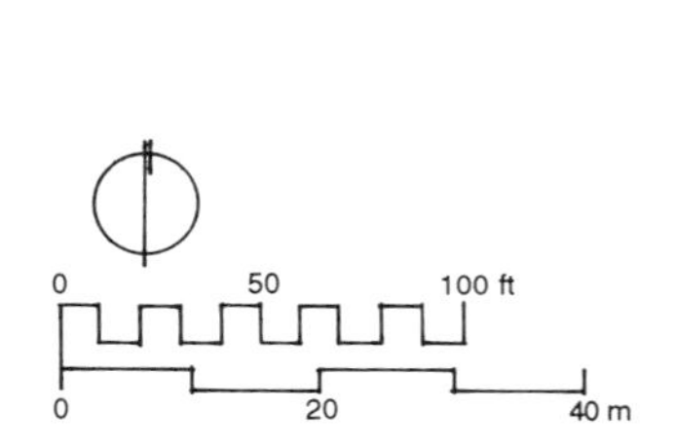

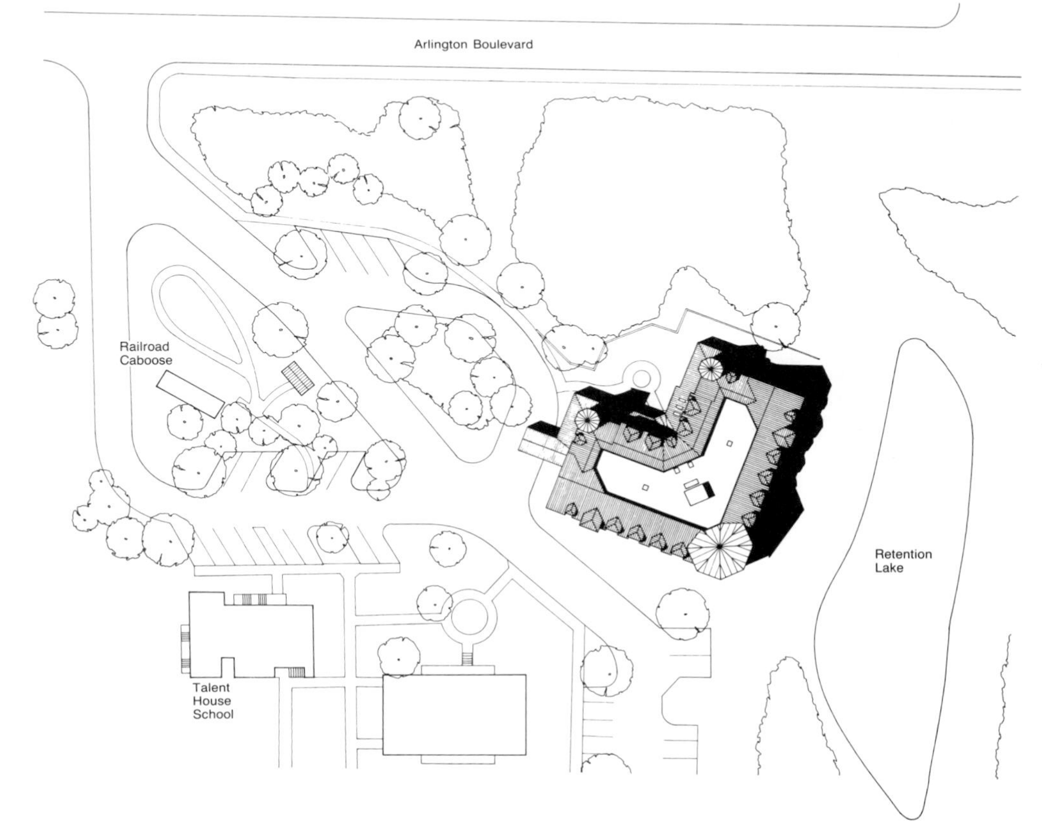

CASE STUDY DIAGRAM 5.22 *Site Plan*: Sited along a sloped edge of a heavily wooded private school campus, *Sunrise* residents have a view of either a lake on the east or a busy street on the north. The proximity of the day-care through sixth grade Talent House School makes a range of intergenerational programs possible. Children visit the *Sunrise* third floor tea room where structured programs with residents are held.

2. Shared Spaces and Common Rooms

Rooms consistent with the residential scale of the building are used for dining and living room functions. The dining room was reduced in size by planning meals around two seatings. A large sculptural open stair centered in the entry lobby leads to the second floor. It reinforces the residential character of the place and provides an incentive for exercise by walking from floor to floor. Skylights throughout the building and tall windows bring added daylight to interior spaces. Crown moldings and a two inch wide chair rail in place of a conventional handrail give the corridor a residential appearance that is rich in detail but avoids the stigma of disability. Short corridors link elevators and stairs to resident rooms.

3. Interior Design Treatments

Many of the interior accessories, casework displays, and the artwork in lounges and on corridor walls are salient to the older people living here. Items include common collectibles from the twenties and thirties. Other furniture items and accessories are displayed in bookcases, on sideboards, or as freestanding objects. These items imbue each space with character, imparting a homey feeling like a resident's living room rather than that of an impersonal entry lounge. Artwork with affective themes features relationships among children, grandparents, and pets, rather than abstract graphics arts. The content of artwork and accessory items stimulates memory and affective response.

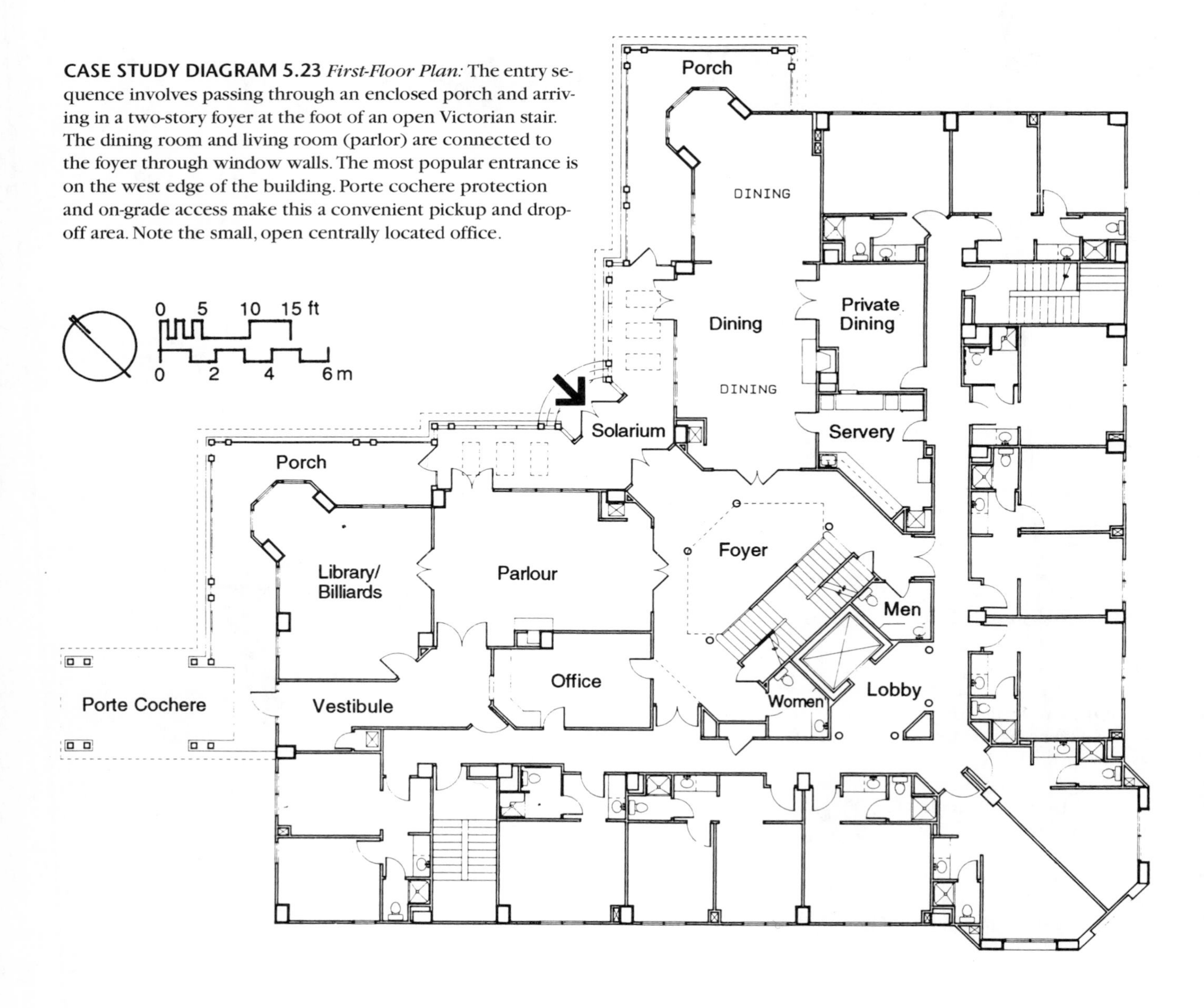

CASE STUDY DIAGRAM 5.23 *First-Floor Plan:* The entry sequence involves passing through an enclosed porch and arriving in a two-story foyer at the foot of an open Victorian stair. The dining room and living room (parlor) are connected to the foyer through window walls. The most popular entrance is on the west edge of the building. Porte cochere protection and on-grade access make this a convenient pickup and drop-off area. Note the small, open centrally located office.

4. Health and Personal Care Services

Weekly checkups record residents' vital signs. Residents with intensive care needs are identified and assisted as necessary. Pets, small mammals, and preschool children are involved in resident programs. Their presence evokes emotions that motivate continued engagement. Programs are designed around the individual interests of small groups of residents rather than a few large management-initiated activities. Call buttons employ a two-tier system that lets residents get minor assistance when needed or summon emergency help if required.

5. Residential Units

Dwelling units are relatively small, but are unique in configuration and offer interesting views of the site. Many special windows or small sitting areas are created by the complex building configuration. Units have no food preparation facilities but a refrigerator allows residents to safely store food and snacks. Closets have been replaced with armoires designed for the reach capacity and storage needs of residents. Residents have keys that unlock their door but rooms are readily accessible to staff if they must enter the unit.

6. Site Planning

A lake on one side of the site and the adjacent preschool on the other, provide excellent views from units. The outside ground plane is separated from the porch and first-floor levels by several steps, making garden areas less convenient and accessible.

7. Changes and Adjustments

There are few informal rooms where residents can meet. A small bistro/café is currently replacing a billiards room so that residents and family members will have an informal place to spend time together. The building's symmetrical L shape makes it confusing for some mentally impaired residents to find their way around. More architectural differentiation at key decision points would help clarify this.

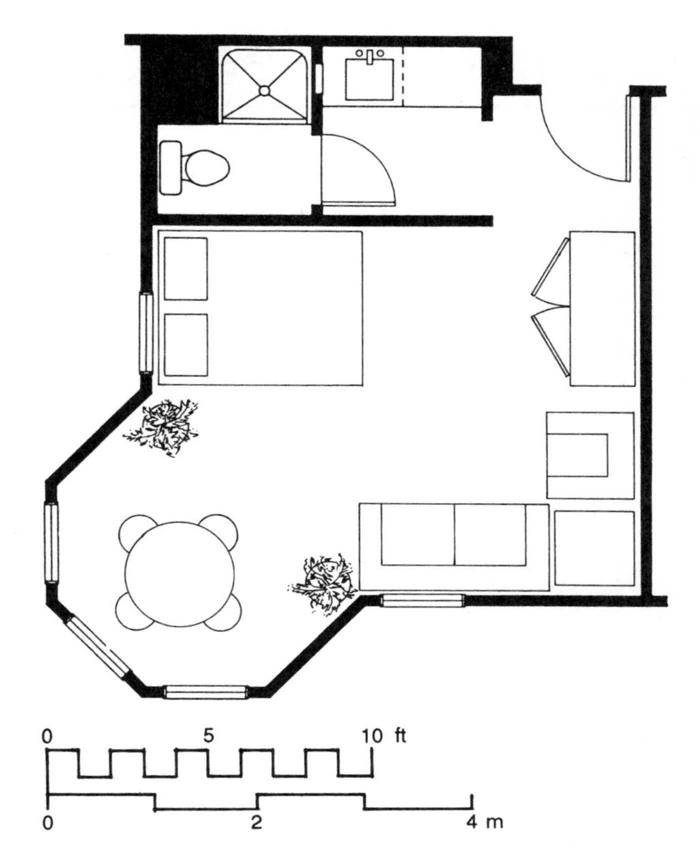

CASE STUDY DIAGRAM 5.24 *Unit Design*: This unit is interesting because it includes a corner turret. Most residents have used this alcove as a living room or have placed a table here for games, writing, and craft activities. Units typically contain a refrigerator and combine bathroom and kitchen functions in one area.

WOODSIDE PLACE, Oakmont, Pennsylvania 7

Architect: Perkins, Eastman and Partners
New York, NY

Sponsor: Presbyterian Association on Aging
Oakmont, PA

Building Characteristics:

a.	# of units	30
b.	# of stories	1
c.	Context	suburban
d.	Housing type	group home
e.	Building parti/shape	courtyard
f.	Unit mixture	6 doubles 24 singles
g.	Size of most common unit (average)	180 sq ft
h.	Community facilities	no
i.	Community-accessible restaurant	no
j.	Year opened	1991

Resident Characteristics:

a.	Average age	80.8 years
b.	# of people	36
c.	# couples/men/women	10 single men 26 single women
d.	% Bathing help	92%
e.	% Toileting help	69%
f.	% Incontinent	28%
g.	% Wheelchair-bound	0%
h.	% Cognitively impaired	100%

CASE STUDY DIAGRAM 5.25 *Axonometric: Woodside* is located between the Presbyterian Medical Center campus and the Oakmont Country Club golf course. The concept of the building evolved from the idea of the Shaker commune. Three residential bungalows on the south edge of the complex each house twelve residents. The northern portion of the complex contains a large "great" room covered by sloping roofs with deep overhanging eaves.

FIGURE 5.13 The sloped roof form of *Woodside Place* communicates a feeling of protection and the imagery of a New England barn. Residential detailing makes it clear this is housing and not a medical care facility.

Summary Description

Woodside Place is designed as a stand-alone, residential, personal care alternative to the typical locked nursing home ward for Alzheimer's residents. It features an open plan with transparencies and half walls between common rooms and a wandering path. This pathway links residents to a range of special-purpose spaces, including a music room, country kitchen, great room, and arts and crafts space. A major strength of the scheme involves the use of three decentralized houses, each of which accommodates twelve residents. These simple bungalow buildings are wrapped around courtyards with a "front door" that opens onto the wandering path. Residential imagery in the exterior massing of the building and in the selection of folk art and Shaker-type finish details inside give Woodside Place the feeling of a rural commune and not an institution.

Project Features

1. Residential Character and Appearance

The individual expression of the three houses on the south edge of the project and the use of a large barn-like sloped roof to shelter common rooms on the north edge give the exterior massing expressiveness, variety, and reference to familiar forms. Inside, handmade quilts mounted as tapestries and Shaker-style casework details remind one of the folk art traditions of western Pennsylvania. Carpeting, wood trim, baseboards, chair rails, and cove treatments give walls, ceilings, and floors a residential feeling. The contemporary art commissioned for Woodside seems curiously out of place in an environment which serves so effectively as a reference to the familiar aspects of traditional residential imagery.

2. Each Bungalow Houses Twelve Residents

Thirty-six residents are housed in three separate bungalow dwellings. Each contains a serving kitchen, laundry area, living/dining room, porch, and courtyard. Staff assigned to each house learn about the special needs and preferences of individual residents. The small scale of each house gives it a residential feel, and increases its manageability. Each house is cued by color and an icon figure (star, schoolhouse, tree) for recognition purposes. Courtyards between houses allow each dwelling unit to be expressed as a separate building in the massing scheme.

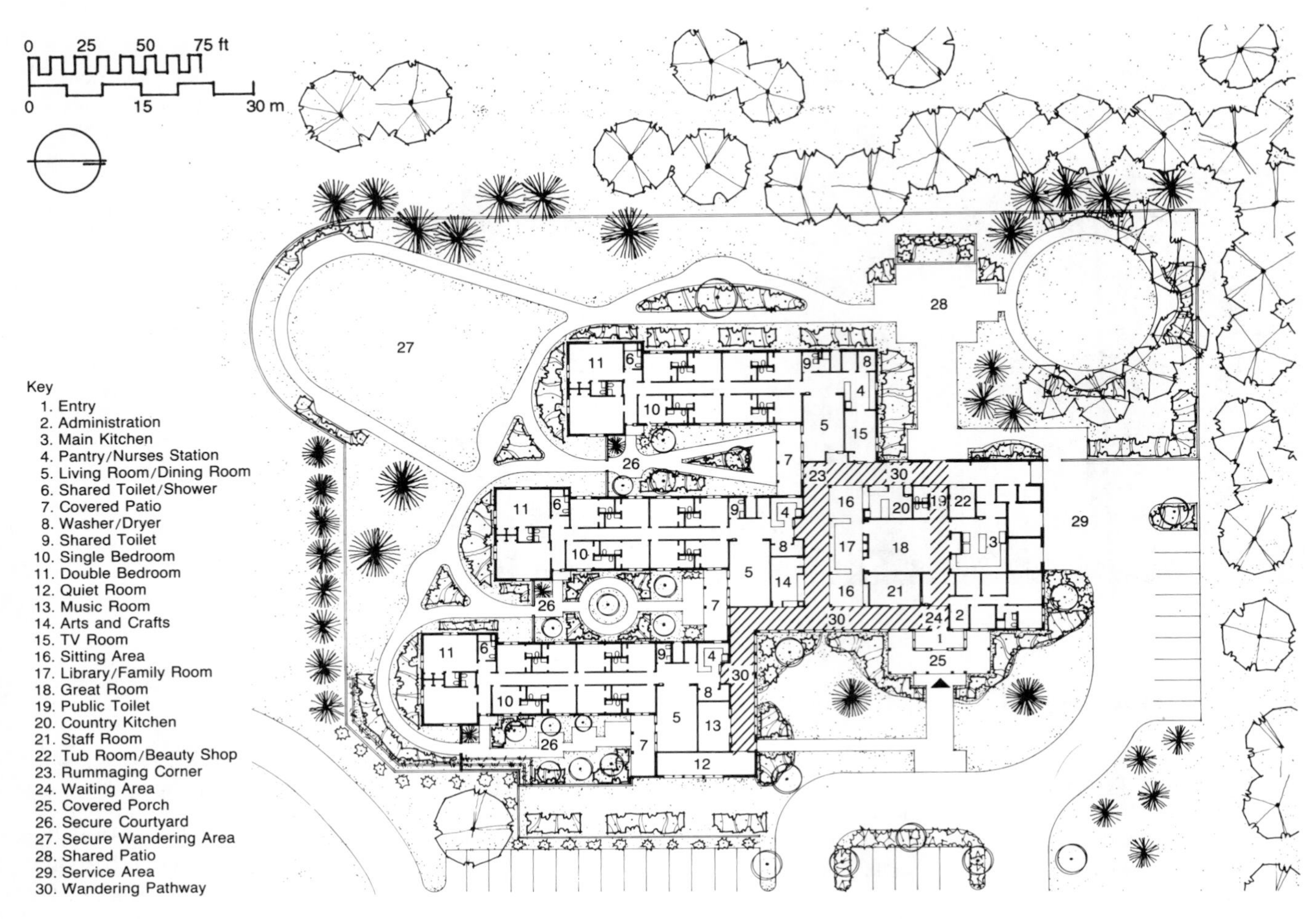

CASE STUDY DIAGRAM 5.26 *Site Plan*: The building plan links interior and exterior spaces by way of porches, corridors, and controlled views. The wandering pathway (darker halftone) plays an important role in the design. Using a city metaphor, the pathway symbolically connects each house to a variety of places where daily activities occur (arts and crafts, library, exercise, music, family visiting, etc.).

3. Wandering Pathway

Woodside is designed to accommodate the aimless wandering which often accompanies dementia. Most conventional long-term care facilities deal uncreatively with this problem by limiting the wandering range of residents to a small area, usually a single linear corridor. The pathway at Woodside is shaped as a large loop that leads residents by each residential house and six special-purpose spaces. Wandering behavior is a major problem that increases the difficulty of caring for dementia victims at home or in an institution. A microchip contained in each resident's name tag activates exterior door locks when residents approach them. Transparent window walls between the wandering pathway and each special use space make it easy for residents to preview and identify activities in rooms attached to the path.

4. Rooms for Activities

Common spaces along the wandering pathway have been assigned specific functions. Each room is designed and furnished to provide families and residents with self-evident cues about how it can be used. For example, the country kitchen resembles a conventional eat-in residential kitchen. The hope is it will stimulate behaviors consistent with the memory of the family kitchen. The music room specified on the basis of its therapeutic value, supports popular and effective group or individual activities. An arts and crafts room is enclosed by glass walls and a large multipurpose space with clerestory windows has been designed with exercise therapy in mind.

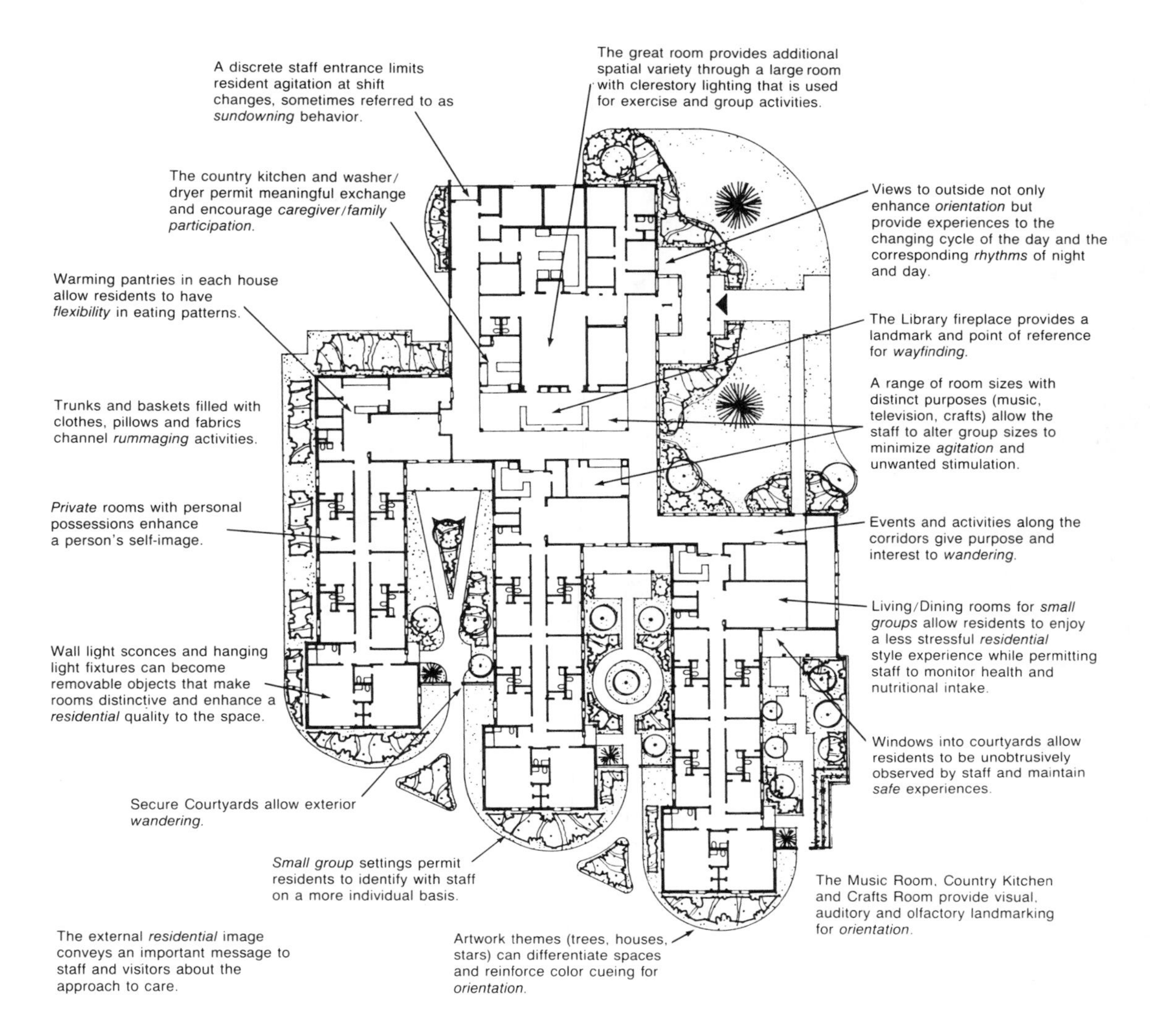

CASE STUDY DIAGRAM 5.27 *Annotated Plan*: This plan describes the intentions behind each major space. The most noteworthy concepts include the decentralization of the resident population into three small houses and the idea of encouraging wandering throughout the facility rather than limiting it to short interior corridors. The complex configuration of the building footprint adds variety to the setting while the looped wandering path allows residents to easily find their way around.

5. Views to the Outside and the Use of Landscape

The complex, varied configuration of the building footprint invites daylight and views to the surrounding landscape. Courtyards are accessible through a covered porch from each house. The use of varied plant material and the unique layout of each courtyard add variety and a sense of individuality to each house. A large landscaped "backyard" lawn with a walking pathway, surrounded by a tall fence, is also accessible. During good weather, residents are free to walk inside and out. Large glass panels that border the wandering pathway provide residents with views of the site and a feeling of freedom.

6. Spatial Linkages and Transparencies

Memory impairments can make the perception of a conventional corridor and room sequence disorienting and confusing. Woodside uses half walls, windows, doors with glass and side lights, built-in counters, and exterior views to cue residents. The door from each house to the wandering path contains a window that allows residents and staff to preview before entering or exiting. Ceiling heights vary according to the size of the room and its need for extra light and space. Some rooms are linked and overlapped while others are self-contained.

FIGURE 5.14 Courtyards adjacent to each bungalow provide attractive controlled views for residents. At one end is a covered porch that connects the living room/dining room with the courtyard. Although each courtyard is the same configuration, the plants and the pavement pattern differ, giving each one a separate identity.

7. Unit Features

Each resident's room is connected to the corridor by a Dutch door. Inside the room, two walls have a continuous three-inch plate shelf for display purposes. A series of pegs below allow personal items to be hung. A wallpaper frieze above the shelf adds variety, color, and pattern to the walls. The original program called for six two-bedrooms and twenty-four private rooms. The two-bedrooms were meant for couples and related individuals. In retrospect, these might have been reduced from six to three. Each house contains a single shared shower. Given the success of models with private shower accommodations in each room, the number of showers might also have been expanded.

8. Entry Corridor

One of the most popular sitting spaces along the wandering path is an area that faces the eastern edge of the site, where the parking lot and building entrance are located. Residents sit here watching the activities of the parking lot, entry, and administrative office. (An entrance designed with benches and an air lock vestibule is inaccessible to residents.) To better accommodate this activity, an enclosed three-season porch might have been designed parallel to the corridor at this point.

9. Discharge Criteria

The design of Woodside Place powerfully demonstrates how architecture can accommodate free and safe wandering in a humane fashion. The building also demonstrates the economics of a less-regulated housing arrangement because it costs less to operate than a conventional nursing home. However, rigid state regulations and self-imposed limitations have kept Woodside from becoming a permanent home to residents placed there. When residents decline to a point where they violate one of eight discharge criteria, they must be moved to a nursing home. The potential exists for Woodside to be not just transitional housing but a permanent home that can accommodate residents as they become more frail.

10. Changes

The living room, dining room, and alcove kitchen of each house could be combined to function more like those in the Swedish group home. Currently, the kitchen design doesn't allow residents to be actively involved in food preparation or serving. Furthermore, the laundry located in a room near the kitchen is awkward and doesn't facilitate therapeutic use. The lounge furniture within the living room seems out of place and isn't used as much as the dining tables, where meals are provided and other activities occur. Visual access from the central administration office to the facility front entrance is blocked, making it difficult to watch residents who are tempted to slip outside.

CAPTAIN ELDRIDGE CONGREGATE HOUSE, 8

Hyannis, Massachusetts

Architect: Barry Korobkin and Eric Jahan Architects
with John Zeisel Ph.D.
Somerville, MA
Donham and Sweeney Architects
of Record
Boston, MA

Sponsor: Barnstable Housing Authority
Hyannis, MA

Building Characteristics:

a. # of units	18
b. # of stories	2
c. Context	small town
d. Housing type	congregate house
e. Building parti/shape	atrium (activity)
f. Unit mixture	2 doubles 16 singles
g. Size of most common unit (average)	275 sq ft
h. Community facilities	no
i. Community-accessible restaurant	no
j. Year opened	1981

Resident Characteristics:

a. Average age	83.2 years
b. # of people	19
c. # couples/men/women	1 couple 5 single men 12 single women
d. % Bathing help	32%
e. % Toileting help	0%
f. % Incontinent	16%
g. % Wheelchair-bound	11%
h. % Cognitively impaired	21%

FIGURE 5.15 The atrium successfully daylights the building interior as well as provides a way to ventilate the building. The elevator and stairs located in the center of the atrium encourage residents to preview spaces before entering them. The *Captain Eldridge Congregate House* has been a powerful and influential precedent for subsequent projects.

Summary Description

Designed as a large remodeled addition to a nineteenth-century sea captain's single-family home on Cape Cod, the Captain Eldridge House has become a powerful precedent for many projects. Its main strength lies in its image as a residential house, which, in spite of its bulk, fits into the surrounding neighborhood well. Inside, a linear skylight over a two-story atrium daylights the center of the building. Balcony overlooks and a central stair landing provide opportunities for previewing common spaces. Units open onto the atrium with double-hung windows and Dutch doors, and through small entry alcoves where plants and personal items are displayed. Common spaces like the breakfast room and parlor on the first floor are designed as comfortably scaled residential rooms. Rich wood detailing gives the interior a homey feeling, which encourages interdependent helping behaviors between residents.

Project Features

1. Residential Imagery

The Captain Eldridge Congregate House was created by adding 8500 square feet to the sea captain's home. The original 2000-square-foot house was restored and integrated into the composition, establishing a stylistic vocabulary for the addition. Materials, details, and the massing of additional square footage were made compatible with adjacent neighborhood structures.

2. Atrium for Natural Daylight

The atrium design creates a strong center to the building, conceptually and visually linking the second floor with common space on the ground floor. During the day, natural light enters the atrium and passes through windows that borrow light from the atrium for dwelling unit kitchens. Thus, daylight enters both sides of a typical kitchen. The atrium assists in natural and mechanical ventilation, adds light to the darkest portion of the building, and increases the attractiveness of the common activities located here.

FIGURE 5.16 The *Captain Eldridge Congregate House* was created through an 8500-square-foot addition to a 2000-square-foot single-family house. The old house was remodeled, and restored. The addition to the building was stylistically consistent with the neighborhood. The exterior fits the building into the neighborhood, while the interior atrium creates a setting for social and communal interaction.

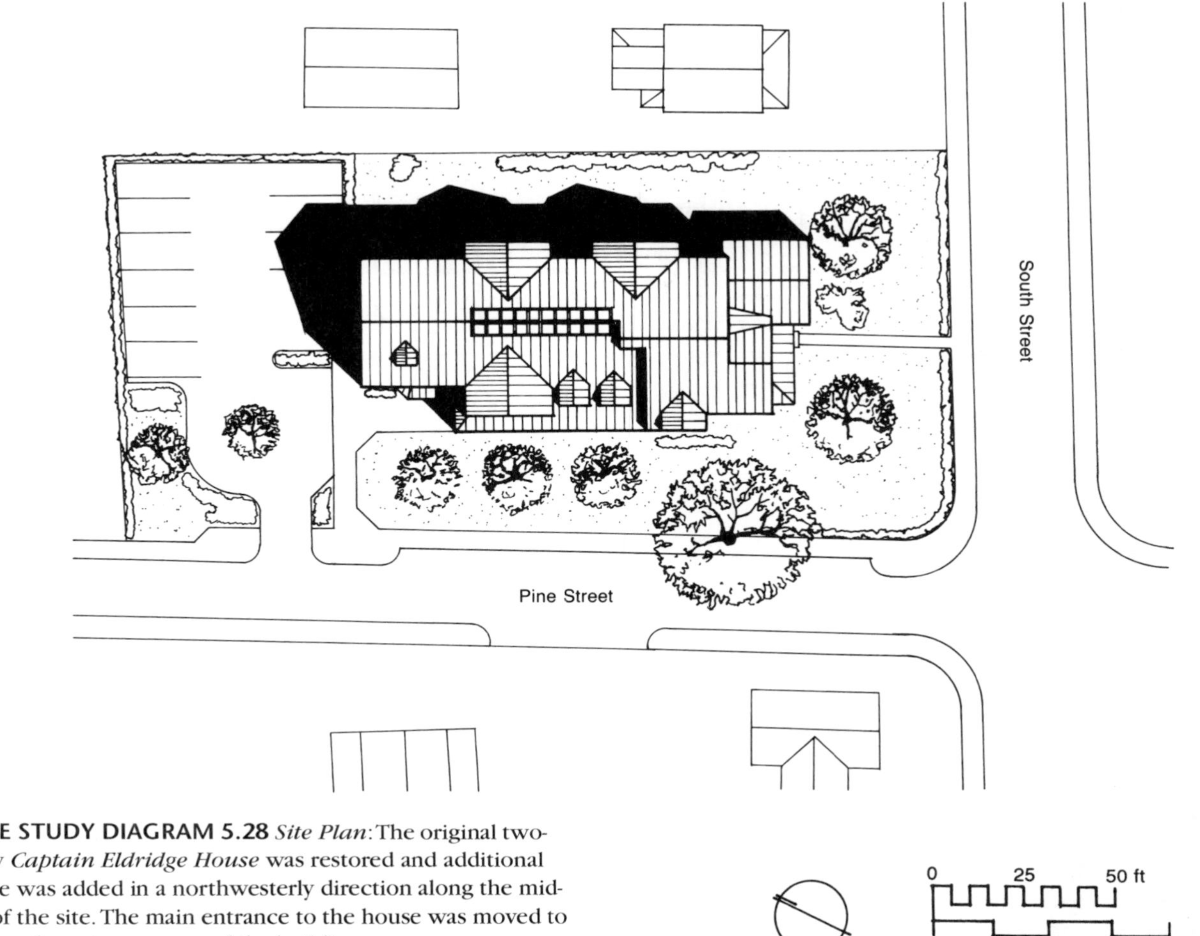

CASE STUDY DIAGRAM 5.28 *Site Plan*: The original two-story *Captain Eldridge House* was restored and additional space was added in a northwesterly direction along the middle of the site. The main entrance to the house was moved to the northwestern corner of the building.

3. Personalized Unit Entries

The entrance to each unit utilizes a Dutch door and a double-hung wooden window to create an identifiable, friendly entry. The half door, a porch light, and the window curtains allow residents to vary the openness of the unit to the atrium, thereby signaling their interest in socializing. The unit entry design also increases opportunities for personal expression.

4. Previewing

Living together in a group setting brings residents into contact with one another in both planned and spontaneous ways. The ability of residents to avoid social contact or make purposeful contacts is facilitated when they can look into (preview) a room before entering it. Stair landings and balcony rails that surround the second-floor atrium allow the previewing of social spaces at the Captain Eldridge Congregate House to occur naturally and spontaneously.

5. Comfortable and Familiar Rooms

Shared spaces are designed around anticipated activities and behaviors common in single-family housing. The breakfast room, adjacent to the kitchen by the front entry, is a popular place for residents to meet for coffee in the morning. The parlor, located in an out-of-the-way area of the original house, is used for special events and formal activities. The outside porch overlooks an adjacent street and can be reached easily from the dining room, before and after dinner. Traditional residential patterns of use were considered in the design, layout, organization, and treatment of shared spaces.

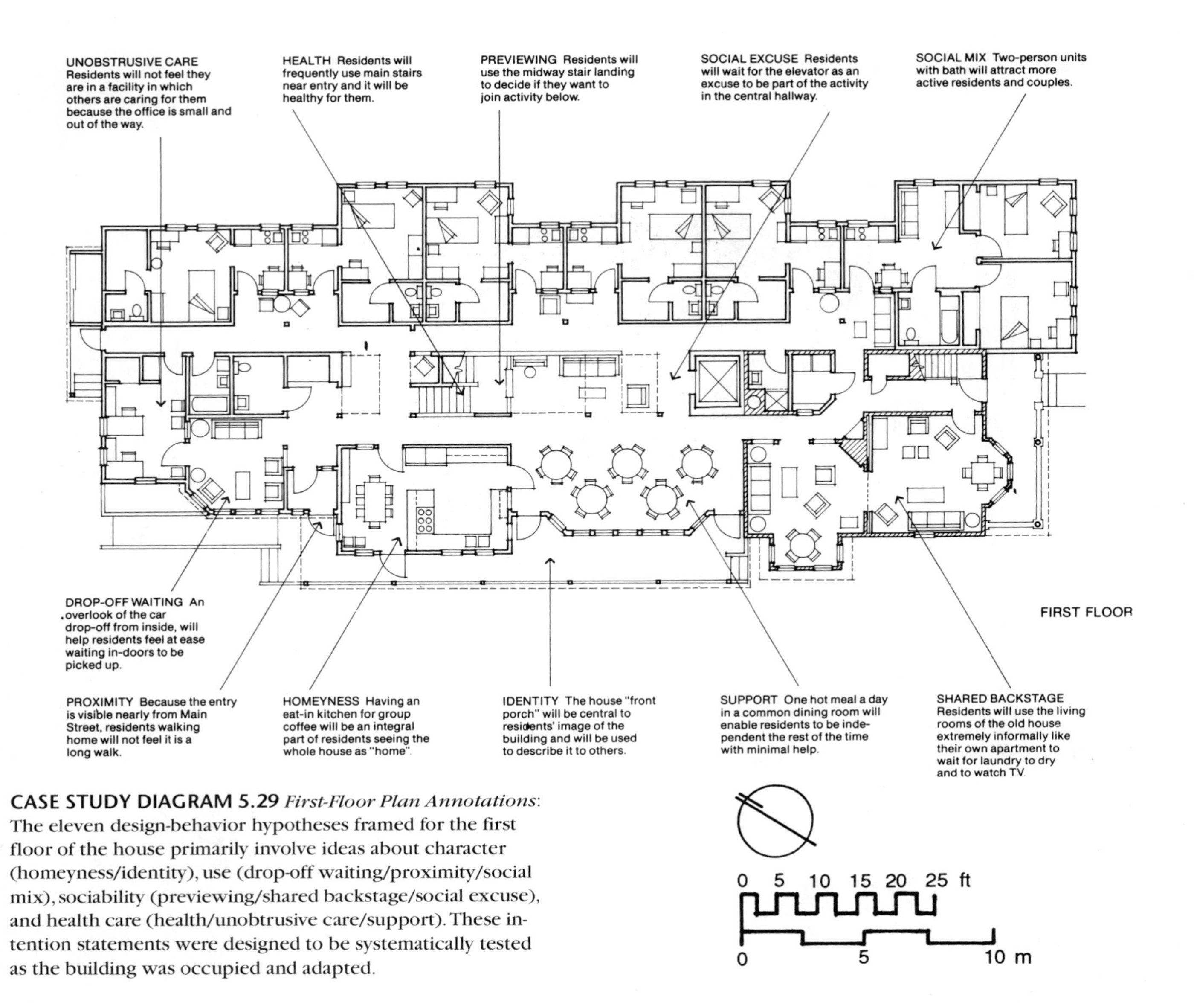

CASE STUDY DIAGRAM 5.29 *First-Floor Plan Annotations*: The eleven design-behavior hypotheses framed for the first floor of the house primarily involve ideas about character (homeyness/identity), use (drop-off waiting/proximity/social mix), sociability (previewing/shared backstage/social excuse), and health care (health/unobtrusive care/support). These intention statements were designed to be systematically tested as the building was occupied and adapted.

6. Open Stair Connection

A centrally placed open stair and elevator connect the first and second floors. The stair is located in a visible and convenient place that encourages residents to use it. It also adds to the comfortable residential imagery of the project. Taking the stairs provides residents with a simple and effective form of exercise. No one has suffered a major injury from slipping and falling on it.

7. Low-Key Management Presence

The manager's office is visible from, but does not overpower, the entry sequence. Its placement epitomizes a management style that encourages interdependence and self-maintenance. Residents operate as a "family," helping one another.

8. Nighttime Care Is a Problem

The size of this housing arrangement and its dependence on externally provided home care services make it difficult to provide residents with night care or to attend to those with confused sleep/wake cycles. This has also limited its ability to retain residents who need more help, so that when additional health services are required, residents must be relocated to a nursing home.

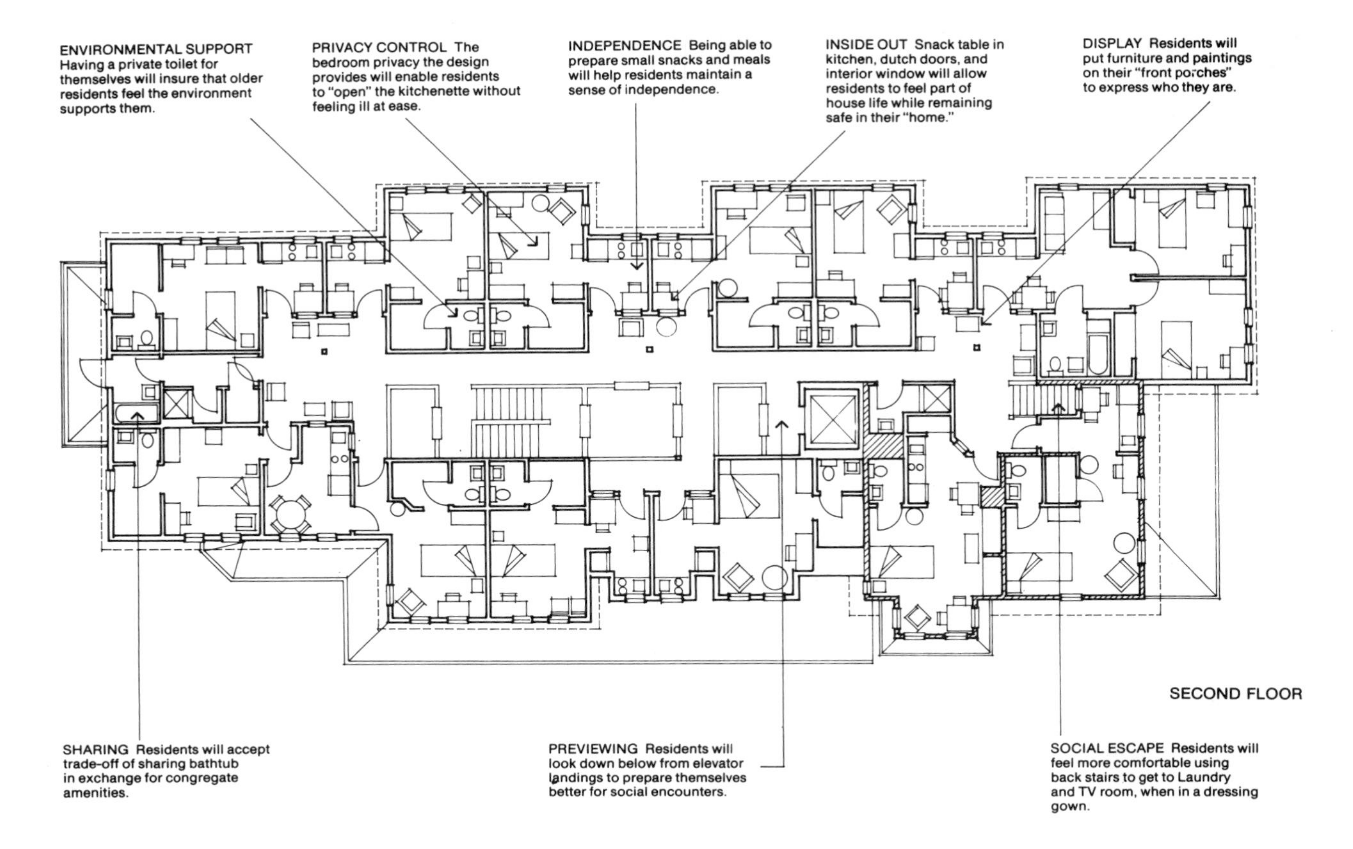

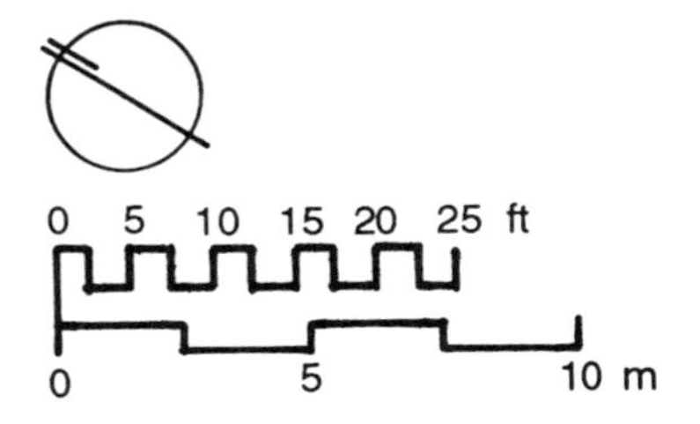

CASE STUDY DIAGRAM 5.30 *Second-Floor Plan Annotations*: The eight design behavior hypotheses associated with the second floor primarily relate to the unit and the visual and physical connections with social spaces below. One of the most intriguing aspects is the interface between the unit and the adjacent circulation area. Dutch doors, window shades, porch lights, and double-hung windows are used to vary the openness of the kitchen to social exchange and connectiveness.

9. Group Bathing Arrangement

Initial construction budget constraints necessitated the installation of shared showers and bathtubs. Management prefers this arrangement because it allows some degree of oversight when residents bathe. Although the house has not experienced occupancy problems, the lack of access to private bathing facilities raises questions about minimum quality expectations as well as its potential impact on future occupancy.

10. Changes

Residents and management would prefer more community space on the upper floor for social interaction. The formal parlor, located some distance from the active central portion of the building, is infrequently used. More office space for home helpers and better storage facilities for supplies would ease the transition to a more intensive service program.

9 LESJATUN, Lesja, Norway

Architect: Hagestands, Stabell og Overehus
Arkitektkontor
Tronheim, Norway

Sponsor: Lesja Kommune
Lesja, Norway

Building Characteristics:

a.	# of units	15
b.	# of stories	1/2 + Bsmt
c.	Context	small town
d.	Housing type	service house
e.	Building parti/shape	cluster
f.	Unit mixture	15 1-bdr
g.	Size of most common unit (average)	490 sq ft
h.	Community facilities	restaurant community room
i.	Community-accessible restaurant	yes
j.	Year opened	1984

Resident Characteristics:

a.	Average age	78 years
b.	# of people	15
c.	# couples/men/women	3 single men 12 single women
d.	% Bathing help	53%
e.	% Toileting help	0%
f.	% Incontinent	27%
g.	% Wheelchair-bound	7%
h.	% Cognitively impaired	0%

Summary Description

This small rural service house contains fifteen units in six small group clusters on a sloping site overlooking a beautiful valley (Lesjatun Serviceboliger, 1990). Buildings are linked together by ramps and walkways, and units are sited with large expanses of glass that face the view to the south. Lesjatun is located adjacent to a nursing home, which facilitates the delivery of home care and health care services. A small 1500-square-foot common building provides meals to residents and older people in the community. Small rural service house models are relatively common in large areas of sparsely populated Norway (Lauvli, 1991). Outside Oslo, Bergen, and Tronheim, most older Norwegians live in small towns or rural villages.

Small towns are experiencing increases in the proportion of older people because of the outmigration of young people. Most small service houses, including Lesjatun, are designed to fit into a village context where individual houses or duplexes represent the majority of the surrounding housing stock. The dominant materials utilized in these dwellings reflect the woodworking tradition of the rural Norwegian countryside. Walls, ceilings, floors, and exterior siding are often made of wood. Special features like carved wooden benches give Lesjatun a strong sense of individual identity and character.

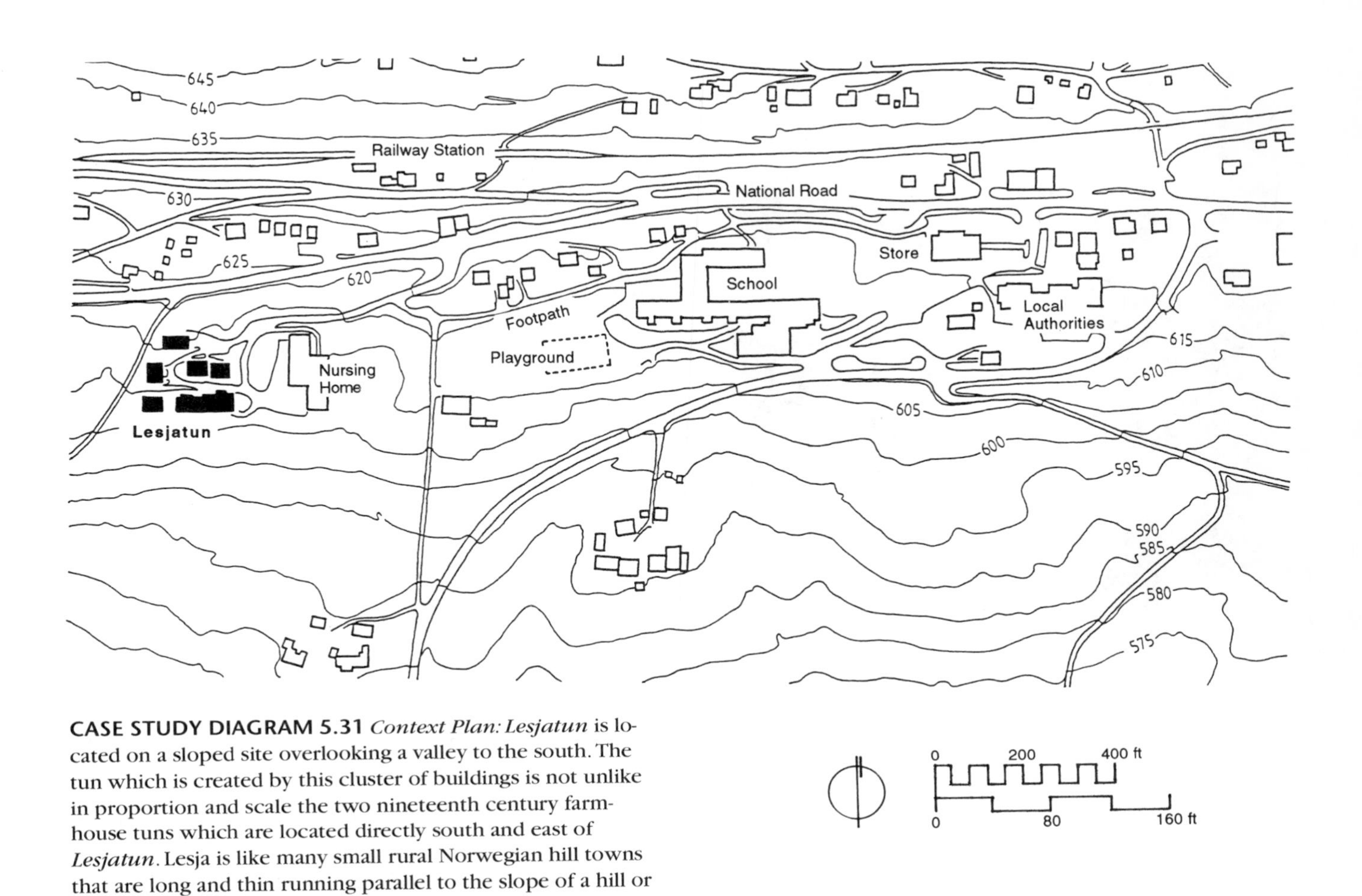

CASE STUDY DIAGRAM 5.31 *Context Plan: Lesjatun* is located on a sloped site overlooking a valley to the south. The tun which is created by this cluster of buildings is not unlike in proportion and scale the two nineteenth century farmhouse tuns which are located directly south and east of *Lesjatun*. Lesja is like many small rural Norwegian hill towns that are long and thin running parallel to the slope of a hill or mountain. (Source: Lesjatun Serviceboliger, 1990; Drawn by Gunnar Selnes)

Project Features

1. Site Location

Lesjatun is located on a south-facing slope, with a view to the south of a beautiful east-to-west-oriented valley. Dwelling units and the service house capitalize on the view and take full advantage of sun exposure. Food, housekeeping, and home care services in the nursing home are located on the same level as common spaces. This facilitates the delivery of food and access to home help personnel.

2. Intimate Residential Scale

Lesjatun consists of six housing clusters that total fifteen dwelling units. The scale of these building clusters is very residential. Only a few two-story buildings are accommodated by the sloping site conditions; the rest are one-story dwellings. The extensive use of wood inside and out gives it a rustic residential feeling.

3. Continuity of Care Is Sometimes a Problem

One of the problems with small projects is their inability to maintain a fully supported night care program. This project, like many other small group home clusters in Norway, has had a difficult time caring for highly debilitated older people who need assistance at night as well as frequent help during the day. However, being next to a nursing home makes it easier to provide some extended care. There has been some controversy in the community over small group cluster housing. Some older people believe the security of a nursing home would be preferable to service dwellings that have difficulty supporting residents on a twenty-four-hour basis.

FIGURE 5.17 The location of Lesjatun on the side of a mountain overlooking a valley that stretches for a hundred miles, is typical of the natural beauty of western Norway. The fifteen units clustered here are adjacent to a nursing home that can provide in-depth support.

4. The Norwegian Tun Links the Cluster Housing Together

The tun in the name Lesjatun refers to the space between buildings. These spaces are used for outdoor activities and have a rich connection to the past. Handcarved benches throughout the tun allow residents to rest as they walk along the hillside from their unit to the common building. In the winter months ramps are slippery. Unfortunately, stairs have not been located alongside ramps to provide residents an alternative.

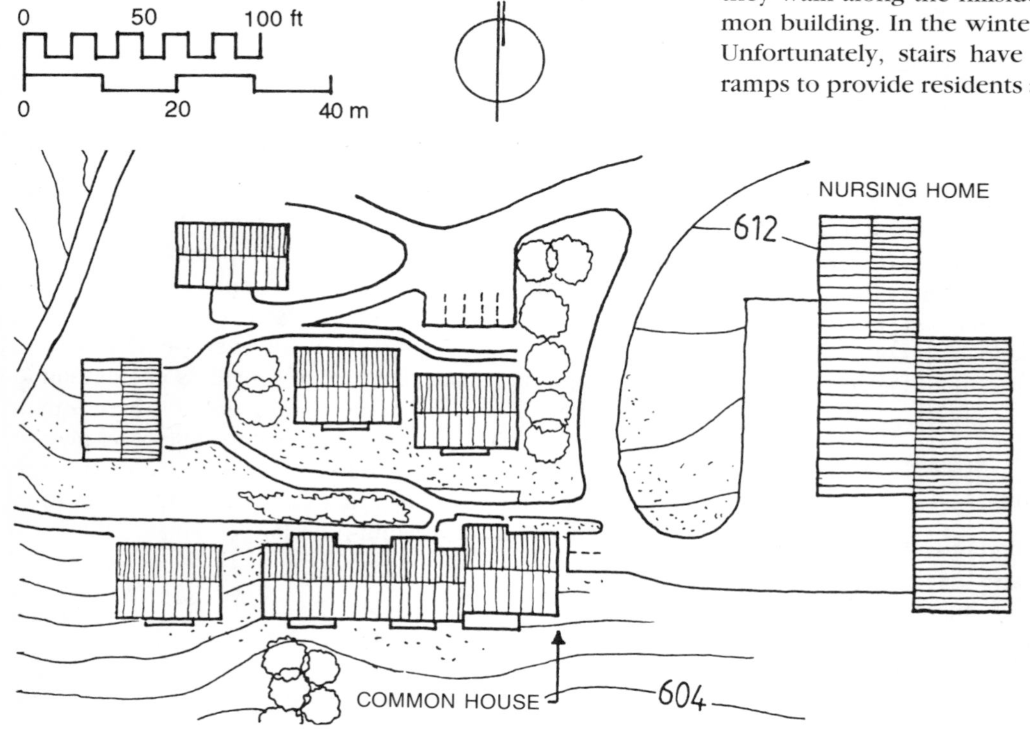

CASE STUDY DIAGRAM 5.32 *Site Plan: Lesjatun* is advantaged by its proximity to a nursing home. Theoretically, 24 hour care could be made available quite conveniently. However, here as in other small scale service houses throughout Norway, the policies that provide supportive care to frail older people in service houses have not been as generous as in other Nordic countries. (Source: Lesjatun Serviceboliger, 1990; Drawn by Gunnar Selnes)

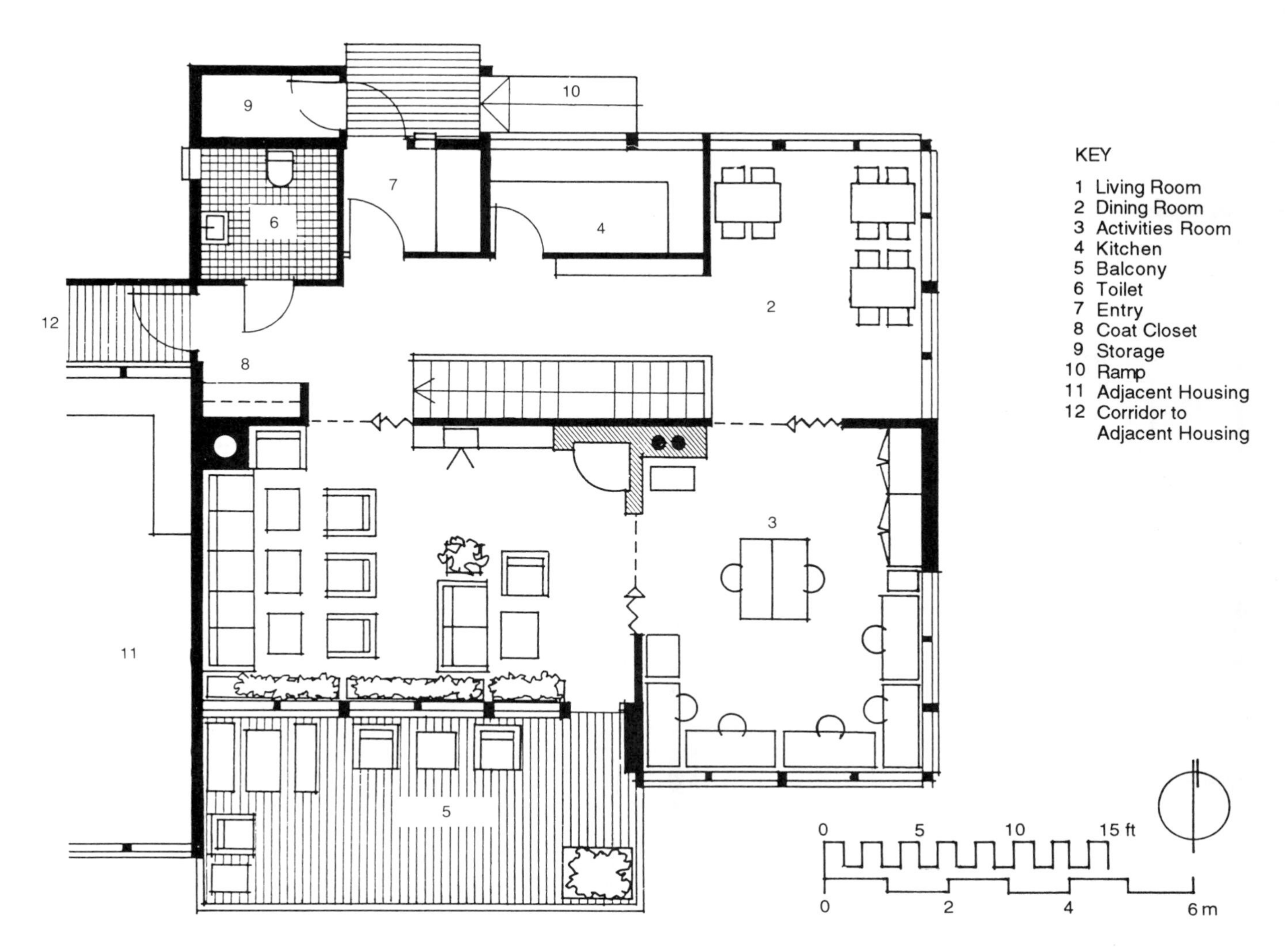

CASE STUDY DIAGRAM 5.33 *Service House Common Space:* The small 1500 square feet service house provides meals activities and social spaces for residents and community members. A two story common building was necessitated by the site slope. The lower floor has a laundry, workshop, and administrator's office. (Source: Lesjatun Serviceboliger, 1990; Drawn by Gunnar Selnes)

5. Service Building Provides Meals and Recreation

Community services are located in a relatively small two-story building with its lower floor nestled into the hillside. Its shape maximizes south sun exposure and the magnificent view of the adjacent valley that slices through this portion of Norway. This service building is open to the surrounding community for meal service, but few people in the neighborhood use it on a regular basis. It also contains a laundry and a workshop. A balcony on the south edge is rarely used and may soon be remodeled into additional indoor space. The living room and craft areas are the most heavily used common spaces.

6. Rural Location Involves Families

One major difference between rural and urban models is the participation of families and community members in visiting. Most families know one another in this community of 2500 people. These widespread social networks make volunteer efforts at the nursing home and the service house more personal and effective.

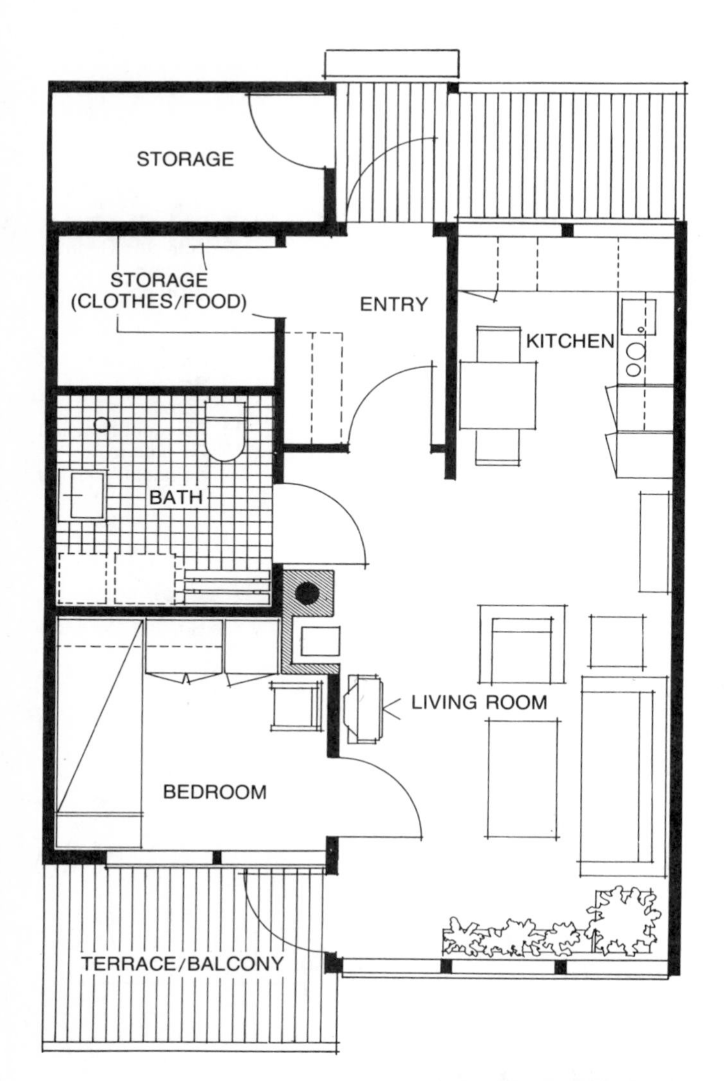

7. Units Are Intimate and Residential in Character

Unit interiors are small but charming, and include woodburning fireplaces. The woodworking tradition visible in the medieval stave churches that dot the countryside continues to influence unit interiors. Wood walls, floors, and ceilings give the unit a rustic homey feeling. Dormer ceilings that slope below eight feet near the balcony edge conserve heat and add to the intimate scale of the environment. Balconies for outdoor activities, large bathrooms for handicapped access, and relatively large L-shaped kitchens are present in the 490-square-foot units. Several units on the south edge of the site have direct access to the service house through a covered walkway. Frail older persons with mobility difficulties often move to this portion of the site.

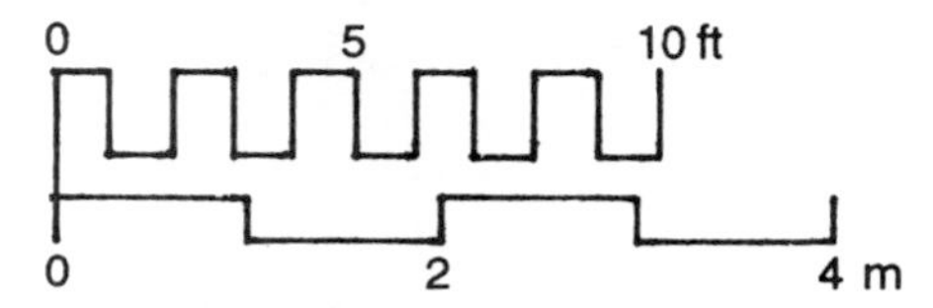

CASE STUDY DIAGRAM 5.34 *Unit Design:* It is interesting to note how space priorities have been established in this unit. The amount of space dedicated to storage is substantial, while the area set aside for the bedroom is minimal. However, although minimal, the bedroom was designed as a separate private room rather than an alcove off the living room. Compared to most US assisted living units, the kitchen is also quite spacious.

FIGURE 5.18 The unit interiors rely on the rustic woodworking traditions of western Norway. Units have roughsawn wood ceilings and walls. Views toward the south invite the sun and open the project to the view of the surrounding valley.

10 HASSELKNUTEN GROUP HOME,

Stenungsund, Sweden

Architect: Arkitektgruppen Strandberg and Partners Stenungsund

Sponsor: Stenungsundshem AB

Building Characteristics:

a. # of units	6
b. # of stories	1 (2-story context)
c. Context	suburban
d. Housing type	mixed use/ group home
e. Building parti/shape	L shape
f. Unit mixture	6 studios
g. Size of most common unit (average)	340 sq ft
h. Community facilities	no
i. Community-accessible restaurant	no
j. Year opened	1988

Resident Characteristics:

a. Average age	85.5 years
b. # of people	6
c. # couples/men/women	1 single man 5 single women
d. % Bathing help	100%
e. % Toileting help	83%
f. % Incontinent	83%
g. % Wheelchair-bound	33%
h. % Cognitively impaired	100%

FIGURE 5.19 The *Hasselknuten* group home in Stenungsund, Sweden, is surrounded on all sides and above by housing for families. The exterior elevation of the building reflects a difference between family units with balconies and elderly units with a canted exterior wall that accommodates living room furniture.

Summary Description

Hasselknuten is typical of the best Swedish group homes for dementia victims, which were initiated eight years ago. This movement developed as an alternative to the institutional psychogeriatric facility. Typically, six dwelling units are planned together as a "family." Staff members eat with residents and everyone takes meals around a large table normally located in the kitchen. The care philosophy involves resident participation in normal daily helping activities. Depending on their competency, they are encouraged to provide help with laundry, setting the dinner table, busing dishes, and shopping for groceries. Attendants also help residents carry out personal hygiene tasks like bathing and toileting. Staff "contact persons," assigned to every resident, know the health histories, family situations, food preferences, and activity interests of their individual charges.

The scale of this housing is small enough to fit into a range of community contexts. Hasselknuten is a purpose-built design located as part of a larger apartment complex that mixes family housing with housing for dementia victims. Some group homes appear more like single-family homes in a neighborhood with detached housing, while others are placed on the upper floors of high-rise housing towers.

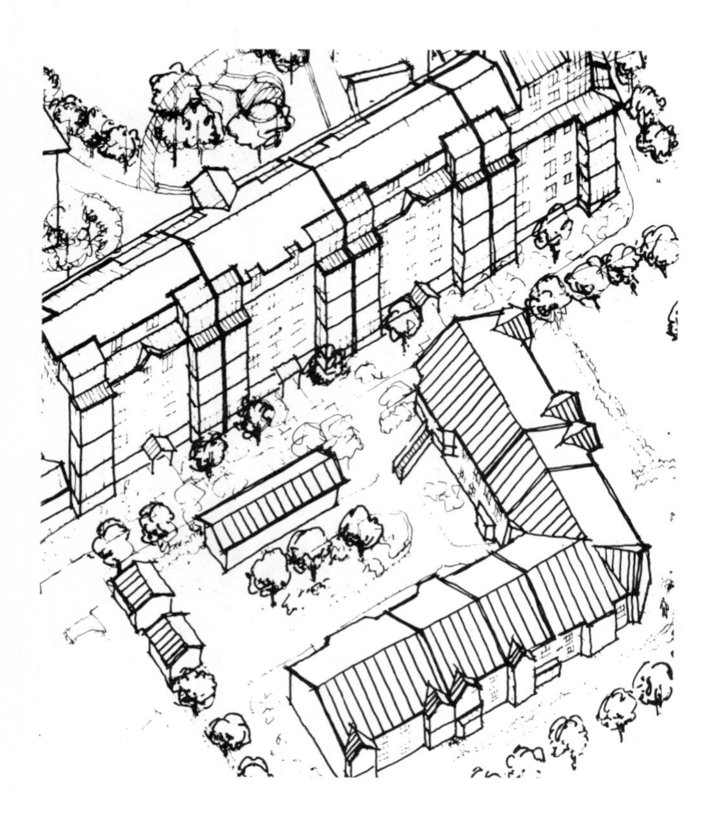

Project Features

1. Residential Imagery

Shared spaces are designed at the same residential scale as a house. In keeping with the homelike theme, dining room and living room floors are wood and a fireplace is located in the living room. Furniture has fabric coverings but they are placed over absorbent pads, which can be removed for washing. Large plants are located in the entry space.

2. L-Shaped Building Configuration

Hasselknuten is designed with two residential wings connected by common rooms and an entry lobby. The entrance to the home is at the inside corner of the L configuration. The entry, living room, dining room, kitchen, and outdoor garden are clustered. When residents are not in their room, they are in one of these spaces. Two large tables are available for group purposes. The kitchen table is used for meals and morning/afternoon coffee. The dining room table is used for special occasions like birthday parties or for recreational activities, including games, puzzles, and card playing.

3. Mixed Use

This group home is part of a larger residential apartment building that contains fifteen multifamily units. It is a two-story L-shaped block. The group home is located on the first floor and is surrounded by family housing on all sides. The family recreation and children's playground is located on one side within the courtyard defined by the building's L shape. The group home has a different exterior appearance than family housing units. Group home units have a bay window alcove, while family units have balconies and patios.

CASE STUDY DIAGRAM 5.35 *Axonometric: Hasselknuten* is in a suburban housing development on the edge of an old historic fishing village. Its L-shape, two-story configuration contains nine family housing units on the second floor and four units at the two ends of the first floor. The group home is located on the first floor and accommodates six memory impaired residents. Visually and conceptually, the group home is mixed with family housing.

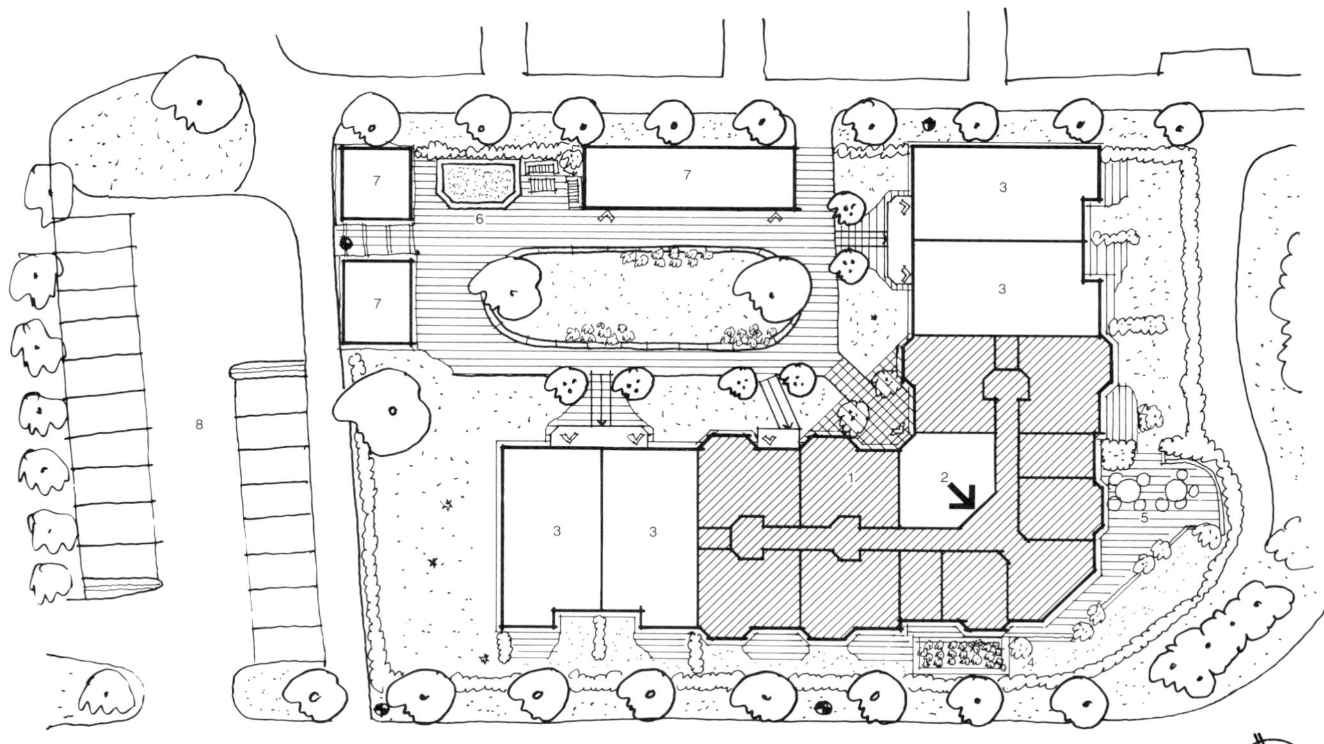

CASE STUDY DIAGRAM 5.36 *Site Plan*: The courtyard formed by the L-shaped building is devoted to circulation and playspace for children. At one end is a sandlot, playground equipment, and a paved area for bikes, trikes, and scooters. A small wandering garden and patio located on the opposite side of the building and accessible from the dining room has been created for dementia residents. Parking is located on the west end of the site.

KEY

1. Group Home (shaded)
2. Public Entry
3. Family Housing Units
4. Dementia Garden
5. Dementia Patio
6. Childrens Play Area
7. Storage
8. Parking

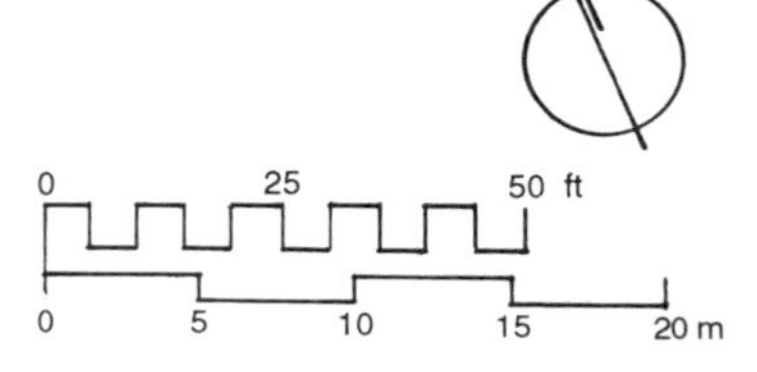

FIGURE 5.20 Transparent glass interior partitions between the kitchen, living room, and dining room make it easier for residents to find their way around. The kitchen is designed to encourage resident participation in food preparation, serving, and cleanup. Staff and residents take meals together, family-style, around a single large table.

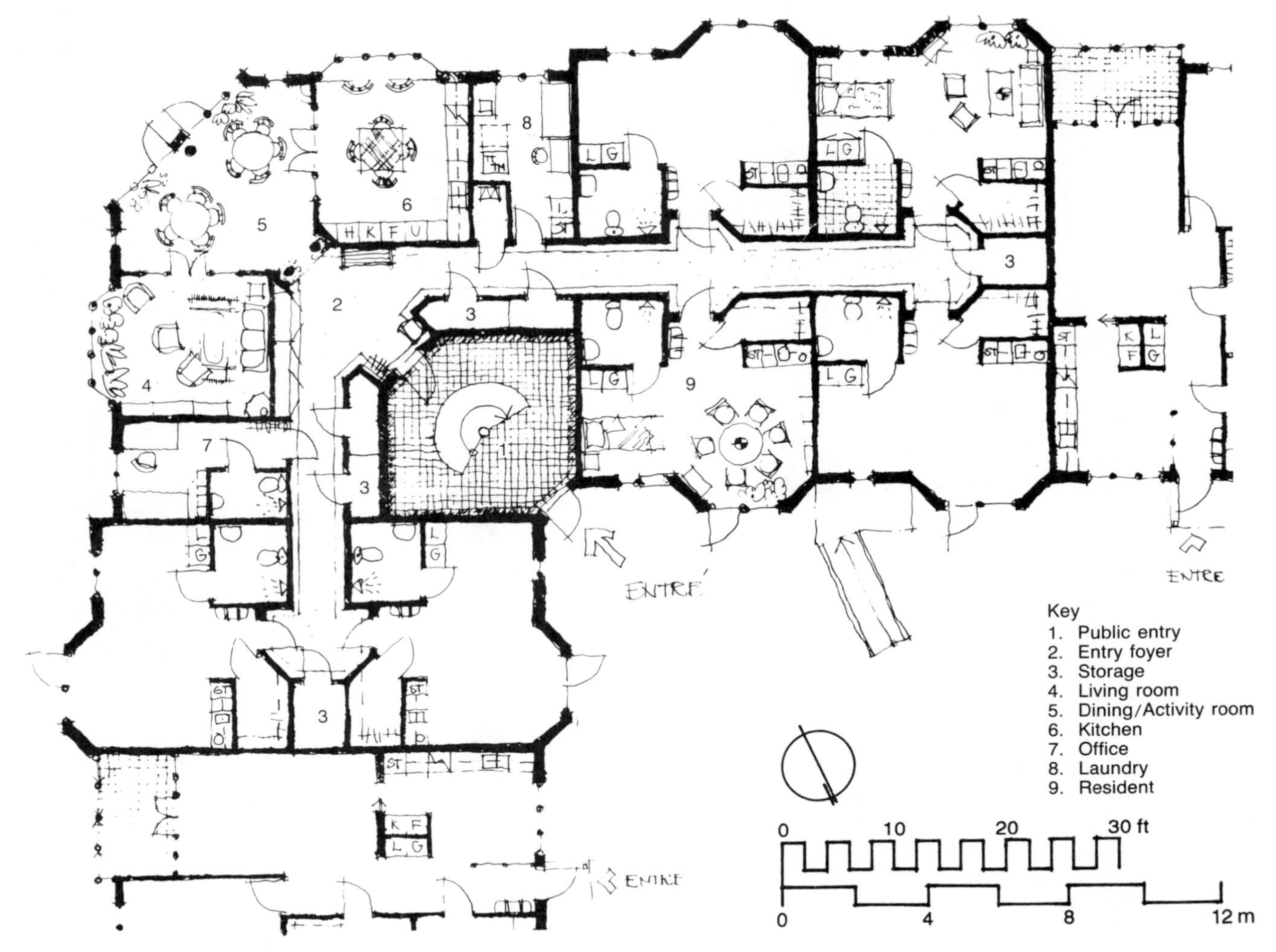

CASE STUDY DIAGRAM 5.37 *Floor Plans*: The group home is entered through a foyer located on the inside corner of the building. A circular stairs here connects to second-floor family units. The L-shaped plan helps to diminish the perceived length of corridors. Common spaces including the living room, dining room, kitchen, and outdoor garden terrace are clustered near the front entry.

4. Kitchen Designed for Therapy

An eat-in kitchen is used to serve meals and engage residents in various normal daily helping activities. Depending on their competency and the food being served, residents are involved in a range of activities, including setting the table, preparing food, and busing, washing, and putting away dishes. The kitchen design facilitates these activities. It is L-shaped with a food preparation alcove on one side and food storage on the other. The dining room table is large enough to accommodate all six residents and the two to three staff members present at meal times. A bay window on one side accommodates a traditional Swedish kitchen bench.

5. Wayfinding Techniques

The L-shaped configuration of the house allows residents to stand in one location and decide which corridor to take. A scale model of one resident's vacation villa is located at the end of one corridor. This example of architectural differentiation creates an effective landmark for wayfinding. Glass walls have been used between the dining room, kitchen, and living room to make the purpose of each room self-evident. Confused residents can see through each window wall and orient themselves by recognizing the activities and people within that space. One blind resident has been given the unit closest to the kitchen to make it easier for her to find the dining and living room areas.

6. Unit Design

The dwelling units are spacious. Each unit contains an alcove for a bed, a full bathroom, a storage area, and a small galley kitchen with a sink, refrigerator, and a one-burner stovetop. The unit is large enough to accommodate a relative for an overnight stay. Residents bring their own furniture and light fixtures.

7. Outdoor Space

A contained outdoor wandering garden is adjacent to and accessible from the dining room. Residents can garden during pleasant weather or just sit outdoors. Part of the garden is devoted to growing rhubarb and berries. Some residents cultivate these plants with the help of attendants.

8. Space for Staff

Group homes also have an office space and lounge where staff can spend time away from residents when they take a break. In some facilities, staff members spend the night when their shift requires them to stay late into the evening.

9. Changes

Many group homes have the same functional problems as Hasselknuten. Major concerns include the insufficient size of laundry and kitchen spaces that do not easily accommodate residents helping with laundry or meal preparation. An additional toilet near the dining room for residents was considered important for those who have a hard time finding their way back to their dwelling unit. A larger staff office was also desired.

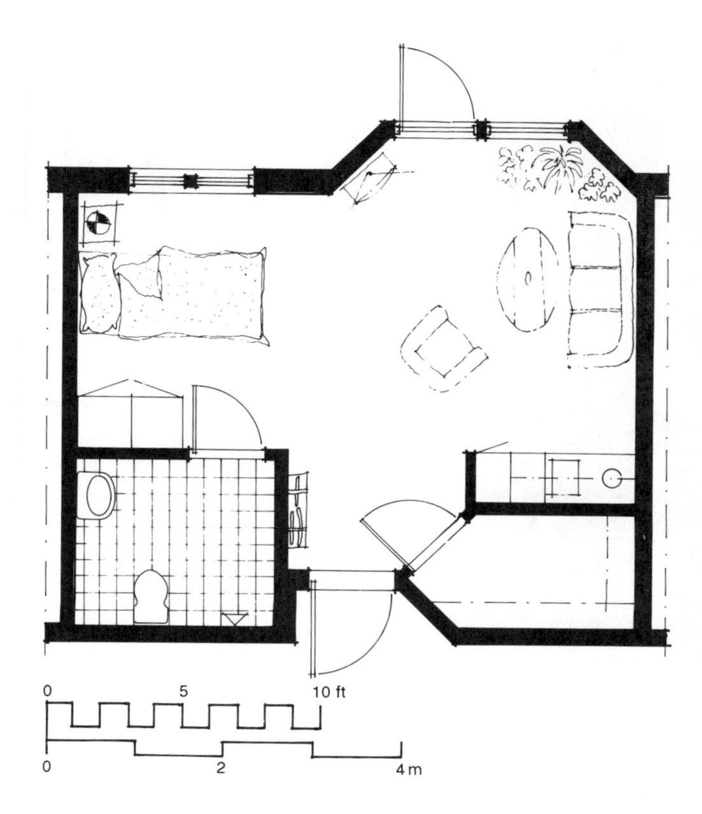

CASE STUDY DIAGRAM 5.38 *Unit Plan*: Each resident has their own private room. The room configuration involves a bedroom alcove adjacent to a full bathroom on one side. The other side is devoted to a storage closet, small strip kitchen with a refrigerator, single burner range and sink, and a living room. Note how the living room has been extended by using a bay window alcove.

FIGURE 5.21 The enclosed street at the *WZV Anholtskamp* project, in Markelo, the Netherlands, clusters unit entries. Linear skylights on the roof allow light to penetrate on both sides of a central bridge. Note the concrete tile floor, which reminds one of an exterior sidewalk.

LESSONS LEARNED FROM DIFFERENT COUNTRIES

While considering the similarities and differences between countries, important contributions from each country became evident. The following summarizes the most intriguing concepts, ideas, and building types from each of the six countries.

Denmark

1. Plegehem

This nursing home environment is designed around a caregiving model that emphasizes rehabilitation through physical and ergotherapy (occupational therapy). Residents typically reside in single-occupied units with a full bathroom. The organization and management approach is decentralized in small clusters of units, with residents engaged in advising management. Residents from each cluster eat together in a small dining room rather than a large group setting. Nurses' stations consist of tables often located in corridors around which residents and staff discuss problems. The decentralized nature of the facility allows clusters to be designed around naturally daylighted single-loaded corridors and landscaped courtyards. The Danish plegehem and the Swedish local nursing home share many similarities.

2. Danish Service House

Similar to the Swedish service house, it has been adapted to a variety of small towns and rural contexts. The Danes emphasize physical and ergotherapy more so than service house models in other Nordic countries. The architecture of these settings is plain in design and simple in execution. The orientation of the service house to both community and project residents makes it an important civic environment. Many projects contain a broad range of special equipment and facilities offered to residents as well as older people in the surrounding neighborhood. Housing attached to the service house is relatively spacious, often in the 600-square-foot range with a full kitchen and bathroom. The curtailing of nursing home construction has meant that service houses now deal with a much frailer population.

3. Co-Housing and Communal Arrangements

Housing designed around communal models of shared care and family support for the frail elderly is relatively rare. A few projects have taken the considerable experience of the Danes in co-housing and translated that into self-help housing models that use facilitators to organize small groups of older frail to care for themselves. In combination with formal home care support services, older residents are able to carry out some of the tasks necessary for collective living, including meal preparation, grocery purchase, and light housework. Many of the popular co-housing models do not serve the older frail well.

The Netherlands

1. Experimental Projects

The Dutch have constructed experimental projects that blend personal care and nursing care into a single residential housing type. Many existing homes for the aged have also been transformed into health care environments through service overlay techniques. Some experiments involve training staff to take responsibility for the medical care needs of frailer residents. Often experiments are structured around benefit-cost assessments. Some of these have been centered on smaller towns where it is not possible to support community facilities, nursing homes, and personal care housing as separate facilities because the demand isn't high enough. A range of new ideas involving cost containment, which challenge and test existing regulatory requirements, is being undertaken.

2. Atrium and Enclosed Street Projects

Out of the twenty-five site-visited projects in the Netherlands, more than half use either enclosed streets or atriums to daylight interiors and connect dwelling units to protected outdoor spaces. Harsh Dutch winters were cited as the main reason for the popularity of this building form. More than 75 percent of the site-visited Dutch projects linked unit interiors with public corridors, enclosed streets, and atria through windows and Dutch doors. Dutch projects, more than others, utilized techniques that personalized the unit entry, making a friendly and highly expressive connection to the corridor. In many projects, artwork or photographs from the older person's house are used to decorate corridors, transforming them from anonymous semipublic spaces to personal extensions of the unit.

3. Self-Help Philosophy

The Dutch are committed to a philosophy of caregiving that stresses self-maintenance and independence. This approach involves encouraging the older person to exercise the maximum amount of personal control in normal daily activities. They refer to this philosophy as providing care by "putting your hands in your pockets" or "sitting on your hands." This caregiving technique often requires more time for staff to assist with tasks than actually doing the task for the older person. However, the Dutch believe that taking self-maintenance responsibilities away from older people makes them more passive and encourages disengagement. Over time, this can lead to "learned helplessness" behaviors as well as muscle strength declines and health problems due to inactivity.

Sweden

1. Group Homes

This radical Swedish movement has provided a viable residential alternative for dementia victims who a few years ago would have been institutionalized in psychogeriatric facilities. Although the group home movement is only eight years old, it is estimated to currently shelter 7000 residents (Lundin and Turner, forthcoming). The idea of the group home is simple. Generally, it is limited to six persons who live within a residential setting that is the size of a large single-family home. In the newest models, residents have their own kitchen and full bathrooms. Attendants assigned to residents involve them in a regimen of daily chores that include tasks such as food preparation, folding laundry, and trips to the grocery store. Evidence from early evaluation studies have shown that residents in these settings do remarkably better than control subjects in institutional environments.

2. Local Nursing Home

This movement, popular in the early 1980s, involved the creation of smaller decentralized nursing homes. Many were designed around courtyards and involved clusters of units where therapy and meals were taken in small groups. Prototypes were developed and the movement influenced the design of many new nursing homes. However, it was brought to an abrupt end when the government declared a moratorium on the construction of new nursing facilities in 1987. This moratorium came about through a shift in old age care from a shelter-based model to an expanded home care model of services delivered to older people living in normal housing in the community.

3. Service Houses

These housing and service arrangements are similar to their Danish counterparts. Housing for the frail is attached to large community service centers that also provide help to surrounding neighborhood residents. Many of these models combine the United States equivalent of assisted living housing, a senior adult day-care center, a restaurant, a home care agency, respite care, and a rehabilitation therapy center.

Norway

1. Rural Service Houses

The majority of Norway's service houses are small-scale arrangements located in sparsely populated rural townships (Lauvli and Guntvedt, 1988; Lauvli, 1991). These projects are miniature versions of urban models that also provide services to residents of the surrounding area. Many of these buildings are constructed of native materials and rely heavily on natural wood for walls, ceilings, and floors. These rustic examples are often organized around courtyard spaces called tuns and are sited to enhance off-site views of the frequently extraordinary surrounding landscape. Although these small centers often encounter problems with the provision of night care, they are frequently located near community hospitals or nursing homes to facilitate the delivery of health care assistance.

2. Life-Span Housing

The Husbanken, Norway's source of public financing, has established accessibility standards that are used to qualify projects for loans (Byggforsk, 1985). The design standards facilitate changes in the environment needed to support older people as they age and become more disabled. Site access, corridor widths, kitchen accessibility standards, and bathroom clearances are examined when plans are submitted for financial approval.

Finland

1. Mixed-Use Responses

Finnish housing projects are often conceptualized as urban additions to the city. As a result, they often employ sophisticated ideas about mixing retail, service, housing, and health care land uses. Finnish projects also include children's day-care centers in ways that allow older residents to overlook playgrounds or engage in volunteer activities. Creative daylighting solutions are often developed that give projects maximum exposure to the minimum amounts of natural light available in the winter. Projects are responsive to the people who are using them and to the elements of the city that surround them.

2. Pool and Sauna Emphasis

The Finnish people have a strong cultural affinity for the sauna and swimming pool. Four of the nine site-visited projects contained a swimming pool and all projects included a sauna. Swimming pools are expensive ventures to build and maintain. Many were approached as joint ventures between cities and nonprofit development corporations that shared these resources with community members. Emphasis on therapy is also present in projects that have formalized movement and physical therapy by placing equipment in visible public places.

FIGURE 5.22 This reproduction of an eighteenth-century tun from a rural Norway farmstead has been reconstructed at an open air museum called Maihaugen, located near Lillehamer. The outdoor space formed by the buildings was used originally to contain farm animals and carry out most of the chores.

United States of America

1. Residential Imagery

The best projects in the United States are experimenting with ideas about residential imagery, which reflect sensitivity to the vernacular tradition of older local housing. Projects have sought to appear as residential as possible in contrast to nursing homes that are characteristically plain and hospital-like, with predictable monotonous configurations. Sloping roofs, porches, dormers, residential window treatments, and friendly entry doors are used to communicate the feeling of home. Porches and arcades also seem to be more frequently employed in United States projects to enhance residential imagery and provide residents with a protected outdoor place to overlook the surrounding landscape.

2. Combining Informal and Formal Systems of Care

Innovative projects have sought ways to develop more effective linkages that allow family members to share caregiving responsibilities. These approaches often involve care management strategies that establish need and employ flexible solutions that blend formal services with informal family support. The best projects stimulate interdependent friendships and informal helping networks among older residents. Families that participate in sharing caregiving tasks with formal providers can save as much as 15 to 40 percent over the costs of traditional nursing homes.

3. Interior Design Treatments

Although European projects benefit from public art funds set aside for commissioned original artwork, exhibited work is often unrelated to the tastes, interests, and expectations of older residents. Interior design treatments in United States projects often involve acquiring furniture and accessories because of their potential affective and intellectual engagement. Interior environments employ artwork and craft items in a therapeutic way that stimulates and engages the older person by triggering long-term memories of positive past associations.

CONCLUSIONS

6

INTRODUCTION

General conclusions are difficult to draw from an investigation like this that focuses on individual case studies with a broadbased research agenda. However, one of the most intriguing aspects of the work was the clear differences in practice and philosophy between European projects and experiences in the United States. The following twenty such differences identify underlying factors, describe how they work and suggest what we can learn from them. Cross-cultural comparisons in the past often began and ended with a discussion of the political, sociocultural, and economic differences between the United States and northern European countries. But the identification of differences in actual practice is less well explored and can provide a better starting point for experimental work.

After all, we are unlikely to reformulate our society along the lines suggested by these heavily socialized welfare states. However, good ideas about physical environments and therapeutic approaches that help older people to live a more independent, satisfying, higher-quality life are completely transferable. We need to understand the European experience from this perspective and learn how to use it to identify limitations and assumptions that quash creative thinking and in the process dehumanize caregiving for the frail.

FIGURE 6.1 At *Woodside Place,* in Oakmont, Pennsylvania, a plate shelf is used by residents to display small personal items. Below the shelf is a series of pegs, which can be used to hang items such as coats, hats, and bedclothing. A wallpaper border above the shelf adds detail and color to each room.

Differences in Practice and Philosophy that Make a Difference

1. Equity and accessibility to all income levels.
2. Commitment to maintain the older person in a residential rather than an institutional environment until death.
3. High-quality housing standards of privacy and completeness.
4. Promoting housing choice in the least restrictive setting based on need and circumstance.
5. Mixed-use and urban design solutions that relate and connect housing to the surrounding context.
6. Service houses that provide help to older neighborhood residents as well as project residents.
7. Supportive housing provided as part of a service continuum where resident needs and preferences dictate placement decisions.
8. Community services that support the needs of family and informal caregivers.
9. Cooperation among public, private and non-profit providers that allow projects to serve a range of constituents.
10. Project concepts that challenge conventional assumptions and existing models.
11. Rules, regulations, and standards explicit enough to ensure quality but flexible enough to encourage innovation.
12. Caregiving professionalism which seeks to meet the needs of families and older residents.
13. Self-maintaining philosophies that encourage resident independence and autonomy.
14. Rural and small-town models that address individual community needs and circumstances.
15. Occupational and physical therapy designed to keep the mentally and physically frail active.
16. Experiments involving communal, cooperative, and co-housing lifestyles.
17. Participatory models of governance involving residents in decision making.
18. Projects of modest size designed to complement the neighborhood.
19. An emphasis on housing form that maximizes opportunities for sociability and observation.
20. A homogeneous population that shares common cultural values.

FIGURE 6.2 The *Fabriken* project, in Jönköping, Sweden, contains a sheltered workshop where gift items are made. Gift items assembled in the sheltered workshop are sold in a shop adjacent to the workshop which faces a heavily trafficked street.

1. Equity and Accessibility to All Income Levels

The combination of national health insurance and ubiquitous welfare policies provide older people of northern Europe with a high level of quality and predictability of services for the aged. Choices are clear and expectations are high. Under most circumstances older people, regardless of income, qualify for services. Housing accommodations are generous in size and service level. The future cost of care is a concern to policymakers and care professionals. In response to this, a range of new approaches to contain costs and increase efficiency is being explored. The vast majority of European projects reviewed in this manuscript are open to low- and moderate-income people. However, due to a lack of public funding in the United States, the majority of assisted living housing projects in this country are targeted toward middle- and upper-income older people.

2. Commitment to Maintain the Older Person in a Residential Rather than an Institutional Environment Until Death

Projects are designed and services organized to keep older people in a familiar residential context for as long as possible. This could be in a service house or a neighborhood apartment, through home-delivered services. Many experiments are focused around providing health services to a range of different housing types. For example, several Dutch experimental housing arrangements are exploring how nursing care and personal care services can coexist freely in the same housing project. Building code regulations do not appear to be as much of an impediment to the concept of aging in place as they are in the United States.

3. High-Quality Housing Standards of Privacy and Completeness

Housing standards vary by country; however, in most northern European countries, current policies advocate the provision of 500–600-square-foot one-bedroom units for residents in service houses. Full kitchens and bathrooms are included in these dwelling units, which are large enough to accommodate guests for socializing or an overnight stay. Services are provided on a case-managed basis. Nursing home standards are also high, with most units single-occupied, 225–250 square feet in size, and outfitted with private toilets and showers.

4. Promoting Housing Choice in the Least-Restrictive Setting Based on Need and Circumstance

Policies regarding housing for the frail in Sweden and Denmark for the last five years have heavily favored services delivered to older people in dwelling units in the community. Older frail people with intensive service needs are often moved to residential dwelling units within a service house. This facilitates the efficient delivery of services and provides a more secure and predictable environment. Institutional placement is considered as the last alternative after all other forms of service-enriched housing have been exhausted. Because of these other alternatives, institutionalization is becoming increasingly less popular.

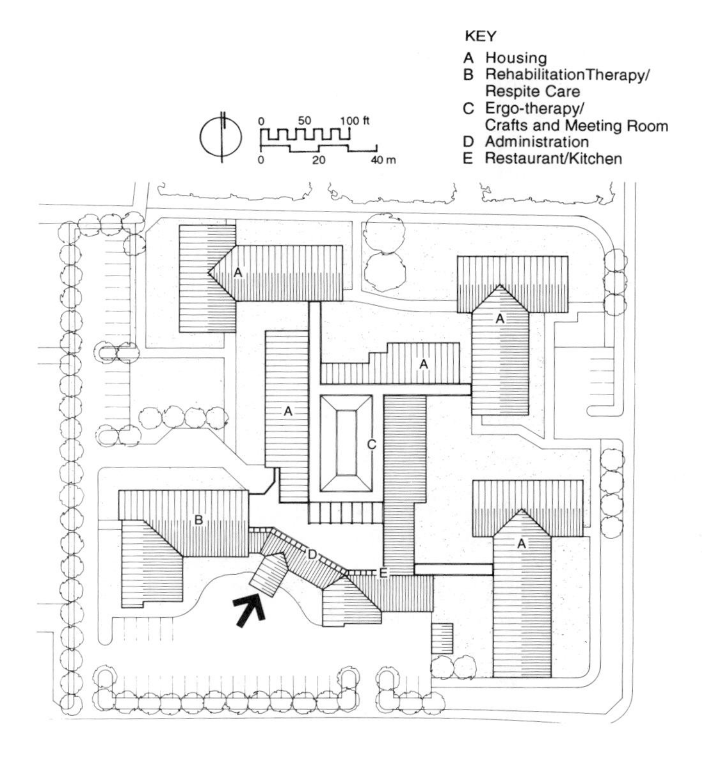

FIGURE 6.3 Service houses combine the functions of housing for the frail with community services in one single central location. *Degneparken,* located in Dianalund, a small Danish town, combines respite care, day care, rehabilitation therapy, a restaurant, meals-on-wheels, ergo-therapy, crafts and recreational activities, home delivered health and personal care services, and housing for the frail. Older people living on-site or in the adjacent neighborhood are welcome to visit for social, recreational and health related services.

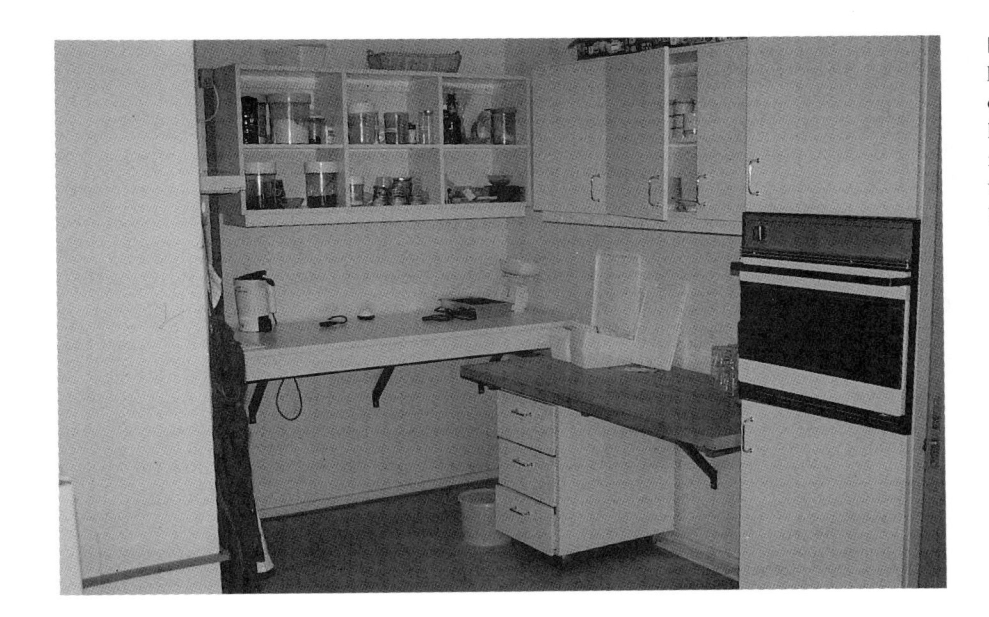

FIGURE 6.4 This training kitchen, utilized by residents of the *Degneparken* service house, is also shared by older neighborhood residents who visit the project on a daily basis.

5. Mixed-Use and Urban Design Solutions that Relate and Connect Housing to the Surrounding Context

Many housing projects mix retail, housing, health care, and community services on one site. Often projects combine housing, health care, physical and rehabilitational therapy, retail goods, home care services, and child-care arrangements in one building. The mixed-use composition of these buildings often connects them more effectively to the surrounding community instead of isolating them. Some projects are urban connectors designed to encourage people to walk through them. Others are located next to important commercial streets flanked by retail services, which make the street edge lively and allow the project to complement the cityscape. The broader constituency attracted to the service house has led to locations that are accessible to transportation and adjacent to shopping districts. In the United States, projects for the frail are more isolated from urban contexts and thus do not encourage community interaction. Furthermore, United States zoning laws make it difficult to mix land uses in creative ways.

6. Service Houses that Provide Help to Older Neighborhood Residents as Well as Project Residents

European service housing is conceptualized in a fundamentally different way than personal care housing in the United States. European projects serve older people in the neighborhood as well as those living in attached housing. Housing is not privatized like it normally is in the United States. It is frequently combined with a range of community services available to older people living in that district of the city. In this way, service houses appear as classic mixed-use arrangements, often combining housing for the frail with a home care agency, senior center, adult day-care center, rehabilitation clinic, restaurant, outpatient health center, special accommodations for dementia victims, and supportive retail services.

The philosophy of the service house is to provide support, therapy, social opportunities, recreational diversion, food, and medical service to older people in need. Operating this way, housing is viewed as a service just like home care or physical therapy, which might be particularly useful and important for an older person, given their circumstances.

7. Supportive Housing Provided as Part of a Service Continuum Where Resident Needs and Preferences Dictate Placement Decisions

Europeans treat housing as one component of a continuum of services offered to older frail people in the community. Older people are moved to a service house when it is considered the best combination of service supports needed to enhance the well-being and independence of the individual. Service apartments are attached to a central facility that caters to the needs of neighborhood residents as well as residents of the housing scheme.

8. Community Services that Support the Needs of Family and Informal Caregivers

Service houses often contain day care, night care, respite care, and rehabilitative services, as well as serving as the coordination base for home delivered services. The service house plays an important role in keeping people independent in their own dwelling unit or within an extended family environment. Northern Europeans have high expectations for elder care services that are based on attitudes about the nature of social entitlements. Community services coordinated through service houses have been very successful in keeping neighborhood residents independent in their own homes.

9. Cooperation Among Public, Private, and Nonprofit Providers that Allows Projects to Serve a Range of Constituents

The commitment of government and nonprofit providers to share resources between residents and the community has led to a number of highly creative approaches to financing. Leased or purchased housing is located in air rights above service centers. In other projects the local municipality owns the restaurant or the swimming pool and leases these common spaces back to housing residents on a time-share basis. During the remainder of the day they are open to community residents of all ages.

10. Project Concepts that Challenge Conventional Assumptions and Existing Models

Project concepts like the group home and the Danish plegehem are constructed around models of care delivery that would be difficult to replicate in the United States because of our laws and regulations. For example, United States nursing homes are designed around compact double-loaded corridors with maximum distances established between a nurses' station and each patient's room. Staffing ratios are linked to the number of nurses' stations that serve a specific population. United States laws also limit the participation of residents in work activities such as housekeeping, laundry, and meal preparation, which is a common therapeutic approach for dealing with dementia victims in Europe.

11. Rules, Regulations, and Standards Explicit Enough to Ensure Quality but Flexible Enough to Encourage Innovation

Regulations are not written or enforced the same way as in the United States. Abuse is considered a minor problem in northern European nursing homes, which are well funded and self-policed. Regulations are written to establish creative guidelines for care providers rather than as a mechanism to police quality. The result is a much more creative and enlightened approach to care delivery, which is not limited to rigid, inflexible, or outmoded rules. New ideas are currently being employed to increase efficiency and shift caregiving responsibility from formal providers to the older person and their family. Although some may argue that the primarily nonprofit and governmental ownership patterns of nursing facilities in northern Europe account for the relaxed nature of their regulatory system, abuse in the United States is not solely concentrated in the private sector. Government organizations such as the Veterans' Administration have been subject to criticism, and nonprofit organizations that operate nursing facilities in the United States abide by the same narrow and stringent rules as proprietary organizations.

12. Caregiving Professionalism Which Seeks to Meet the Needs of Families and Older Residents

Respect given to care professionals seems higher in northern Europe than in the United States. This could be partly explained by the nature of the northern European regulatory system, which is based more on performance measures than adherence to specific rules and protocols. In the United States, regulations tend to be prescriptive in nature, leaving little to the imagination and discretion of providers. In fact, when effective and useful caregiving strategies violate the specifics of the law, providers are more likely to be punished than rewarded for seeking a creative solution to a particular problem. This fear based system discourages professionalism while eroding the self-esteem and creative problem solving spirit of caregiving personnel.

13 Self-Maintaining Philosophies that Encourage Resident Independence and Autonomy

Providing care by "sitting on your hands" or by "placing your hands in your pockets" literally translates into philosophies that engage older people in self-maintaining activities that encourage them to accomplish as much as they can on their own. This philosophy is consistent with the techniques of managed risk and case management assessment that establish a care plan for each resident. It assumes the competency level of an older person can be improved by building existing abilities and restoring lost abilities.

14. Rural and Small-Town Models that Address Individual Community Needs and Circumstances

In small towns, the demand for nursing care and assisted living is often not enough to justify the development of two separate facilities. Hybrid models that mix residents of varying competency in service apartments permit a facility to achieve economies of scale while allowing residents the opportunity to exercise maximum independence for long periods of time. The small service houses of rural Norway have home care programs that cater to the needs of residents in the community, thereby expanding their influence. Buildings that contain flexible meeting spaces, which can be used by other community organizations for a range of other intergenerational purposes, allow these capital improvements to have broader utility within the community. The ability of private and nonprofit sponsors to mix the public needs of the community with the housing and service needs of the frail can satisfy a broad range of community needs. United States projects could profit from such mixing of public and private resources.

15. Occupational and Physical Therapy Designed to Keep the Mentally and Physically Frail Active

The Danes, perhaps more than others, have developed a rigorous approach to therapy that views physical and occupational activities as an important strategy to reduce "medicalizing" impacts and replace sedentary behaviors with an active lifestyle. Residents are expected to engage in mental and physical exercise because it promotes self-esteem, maintains existing abilities, and encourages greater independence within care settings. The Danes view these forms of activity as central to a successful life. Thus, physical and ergotherapy take on symbolic meaning and purpose. Facilities are not viewed as warehouses or places to die, but as therapeutic environments that help residents lead a more successful and fulfilling life.

16. Experiments Involving Communal, Cooperative, and Co-Housing Lifestyles

Most northern European cultures have experience with collective living in a range of different ways. In one creative experimental project, a facilitator manages and assigns work activities to frail residents. Residents contribute something to household management according to their ability, while tasks that are beyond their ability are handled by professional caregivers.

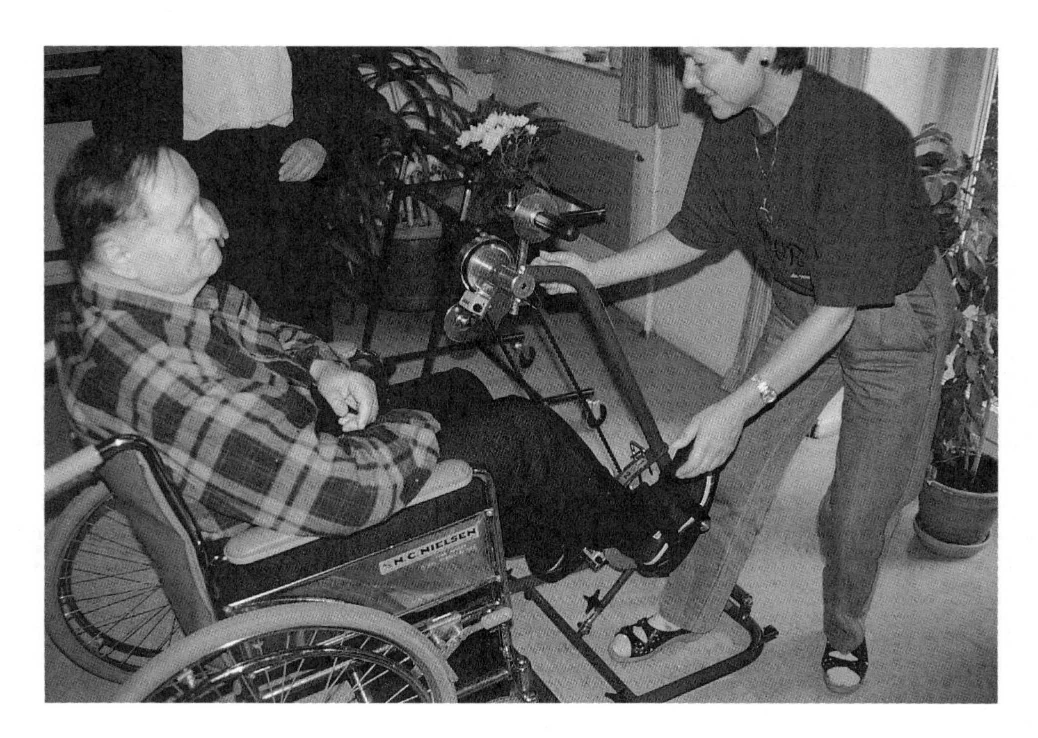

FIGURE 6.5 In the Danish nursing home, every resident is assigned a therapy regimen. The attitude toward physical therapy is so pervasive, it appears to be a lifestyle attribute. Residents are expected to perform at their highest level of competence to minimize their dependence on staff and maximize their independence and autonomy.

FIGURE 6.6 The *Old Peoples Home and Health Center,* at Oitti, Finland, was designed in response to an architectural competition. In order to link an adjacent neighborhood to the town, a tunnel connection was devised through the housing project. The lounge space pictured on the left overlooks this active sidewalk (see Drawing 4.5, View #6). From above, residents can watch children bicycle and adults walk through the project.

17. Participatory Models of Governance Involving Residents in Decision Making

Many Danish nursing homes are organized around a decentralized management model that allows residents and staff at a ward level (fifteen to twenty beds) to meet and discuss issues related to caregiving and staff hiring. Meetings occur as often as twice monthly, and are reported to be both lively and instructive. Residents participate and add to the staff's general level of understanding of their needs, giving residents a better sense of control over their lives. Resident participation works because of the small decentralized size of unit clusters and the discretion given to team leaders at this level. This approach allows management to be more sensitive to the needs and preferences of residents.

18. Projects of Modest Size Designed to Fit and Complement the Neighborhood

During the past ten years, northern European housing for the frail elderly has moved away from large-scale projects to smaller-scale developments that often fit the community better, and are less threatening and overwhelming to neighbors. Many of the newer projects are in the thirty- to seventy-unit range. When providing services to the surrounding community, it is easier to design housing that is smaller in scale because economies of scale in service production are achieved by jointly serving neighborhood residents. The regulations that govern northern European arrangements also do not penalize smaller aggregations of units.

19. An Emphasis on Housing Forms that Maximize Opportunities for Sociability and Observation

Numerous new ideas have focused on making housing arrangements more responsive to the social and behavioral needs of residents. Clustering unit entries and linking units to corridors with windows increases the perceived friendliness of the setting and the potential for social exchange between residents in common spaces. Atrium projects define and create accessible and protected spaces for interaction and recreational activities.

20. Homogeneous Population that Currently Shares Common Cultural Values

The homogeneity of the current population in northern European countries and concerns over equity for fellow countrymen have made it easier to implement strongly supported care programs for the disabled, handicapped, aged, and children. However, many of these countries are experiencing large influxes of immigrants from places like north Africa and Turkey. In the future, the homogeneity that has characterized these societies will not be as great. Currently, ethnic immigrants are inclined toward family-based care that involves parents and grandparents in three- and four-generational families. Larger numbers of immigrants and their adaptation to northern European cultures will mean that more ethnically diverse populations will likely enter the formal care system in the future. Attitudes concerning access of services to minorities may become more divisive in the future.

SUMMARY CONCLUSIONS

In reviewing the best projects from northern Europe and the United States, the following five issues struck me as some of the most powerful lessons to be learned from this experience. I believe they have important implications for the way we design, organize, and provide supportive housing for the mentally and physically frail in our society.

ONE: Models of Community Care

The northern Europeans have designed a system that mixes community services with housing for the frail. They have done this by developing mixed-use arrangements, by building in air rights above other land uses, and by establishing the service house as a concept for integrated service provision within the community. Providing an extensive array of services to older people living in the surrounding neighborhood allows them to anticipate when community residents can profit from a move to the service house. Individuals are kept independent in the neighborhood for as long as possible. When they need the security of a twenty-four-hour arrangement or require intensive services on an ongoing basis, they can be assessed and moved.

Models of community care work in conjunction with housing by treating it as a service that can be prescribed when necessary. In this way, the service house does not have the stigma that we associate with an institution. The service house is a place, not only for older sick people, but for receiving help and assistance in remaining independent. This gives it an important civic function in the community, further reinforcing its identity.

TWO: Interdependence, Self-Maintenance, and Family Support

The basic philosophy of providers is to encourage older people to do as much as they can for themselves. Frail residents are challenged to maintain their competency at its highest level. Nurses and service workers serve by encouraging them to do things for themselves, rather than by doing everything for them. This attitude suggests services should be provided in a therapeutic way that builds and maintains resident competency.

In the United States, we are recognizing that one way to reduce health care costs is to create partnerships with formal care providers that allow families to continue to care for residents by sharing caregiving responsibilities. This idea has been popularized through home care assessment models. The major benefit of these plans is the continued engagement of the family in the life of the older person. We are also recognizing how older residents can maintain their independence by helping others and staying actively involved in normal daily routines. Helping one another provides a sense of contribution that bolsters self-esteem and minimizes formal intervention.

THREE: Autonomy and Privacy for the Rich and Poor

The standard for housing for the frail is changing throughout the world, from one based on a hospital model epitomized by the nursing home to an idea that involves delivering health and personal care services to older people in a normal residential environment. This transition involves a fundamental shift in thinking, which balances the traditional residential qualities of privacy, autonomy, personalization, and control and choice, with the concept of safety and oversight provided by an institution. The idea of managing risk within this type of setting brings into natural balance the desire for safety and the need for autonomy.

A disturbing difference between the European projects visited and exemplary projects identified in the United States is the clear lack of public support for American models. Of the 100 site-visited European projects, 90 were designed for moderate- or low-income populations. In the second round of expert-nominated American assisted living projects (which netted 100 projects), only 10 were designed for moderate- or low-income populations. We have abandoned the older frail person by providing public access only to nursing home care rather than thinking about our obligation to keep the frail out of this setting for as long as possible.

FOUR: Respect for the Professional Caregiver

You can understand a lot about the quality of care provided in a typical facility by listening to how older people are addressed by staff. I often count how many times I hear something said that an older person finds amusing. You can also sense a dedication and concern for helping that makes you feel good and creates optimism and delight where despair and depression are often subliminally omnipresent. Here the physical environment makes only a meager contribution. It is the staff, their attitudes, and their dedication to the work which makes the difference. As a society, we have not given this high priority, while northern European societies have degree-granting institutions and training programs that support professional approaches to care management. These societies treat caregivers for the aged with the same respect that school teachers received decades ago. Both are important jobs that any self-respecting, civilized, humane society should hold in the highest regard and do much more to support. (In northern Europe, the level of creativity and the autonomy assumed by the lowest-level service worker is accountable to an ethical set of professional standards.)

In the United States, we have created a long-term care system that is fear-based. It assumes incompetency and violation of standards is the norm rather than recognizing professionalism and reinforcing it. We police facilities citing them for violations of standards even when these standards are not consistent with the needs, desires, and interests of older residents and their families. We have narrowed possibilities in an effort to minimize problems and as a result have thrown out positive and inventive ideas along with bad practices. We need to support professionalism, creativity, and responsibility rather than establish a set of narrow predetermined rules that denigrate and dehumanize the caregiver and the older person.

FIVE: Systems of Regulations that Encourage Innovation, Experimentation, and Cost Containment

Most new European housing models dealing with the needs of the chronically mentally and physically frail older person are experimental. Each building is viewed as a potential experiment for testing new ideas and best practices. Many of these buildings are formally evaluated and information about success and failures flow back into the system to inform practice the next time a building is initiated.

Furthermore, regulations and standards, which could be viewed as stringent even by United States standards, are flexible enough to be challenged, interpreted, and pursued in a variety of different ways. Europeans are experiencing the same economic malaise as the United States and must deal with a fixed sum of resources. In developing new models, two major criteria are applied to all new project designs.

1. Will the new idea contain costs more effectively?
2. Will it provide a higher level of residential satisfaction to residents?

If a preliminary idea meets these two criteria, then it is deemed worthy of further exploration.

Many of the ideas being actively explored in European projects are based on techniques and ideas precluded by United States regulations, and if pursued in this country they would be considered illegal. We need to examine the purpose and rationale behind building codes, zoning codes, licensing requirements, and state-mandated regulations. Instead of trying to anticipate the myriad of problems and possibilities that exist, we should establish a performance standard which encourages innovation and serves the best interest of residents and their families.

FIGURE 6.7 Volunteers at the *de Kortenaer* project, in Helmond, the Netherlands, assist the staff in conducting the social program. These volunteers live in the project or in the surrounding neighborhood. This housing arrangement is designed to be transformed into a more service-intensive environment as residents age and need more formal assistance.

Concluding Remarks

Change is the issue that underlies most concerns raised in this chapter and in this book. We are perpetuating a long-term care system that doesn't work. Policymakers consider it too expensive, consumers consider it substandard and inhumane, and family members feel alienated and frightened by it. It is a system that costs too much and delivers too little.

This book has focused on how we can view this problem from a new "outside" perspective. We have operated by digging ruts into a conventional landscape that are so deep they keep us from seeing over the horizon that is the status quo, and so steep that they seem impossible to overcome. The European point of view raised in this chapter is not inconsistent with the best housing projects we identified and analyzed in the United States. We can make these ideas work in the United States or we can create a way to develop our own future, which learns from these experiences and improves what we are doing.

I hope this book has given you ideas, insights, courage, and resolve to question many of the inappropriate and counterproductive aspects present in our system of long-term care. Making a better future for society in this regard should be both self-serving and dedicated to the highest ideals of a compassionate society. This is a problem that won't go away and won't improve unless we are all willing to commit ourselves to the difficult challenges that lie ahead.

APPENDIX A

Noteworthy Projects

INTRODUCTION

The following contains names, addresses, and noteworthy features of projects referenced in this book. In completing the research for this book, 100 European and 25 United States facilities were visited. The following list is reduced to the 100 best examples. Included are sixteen from Denmark, twenty-five from Sweden, nine from Finland, eleven from Norway, twenty-four from the Netherlands, and fifteen from the United States.

These buildings have made contributions to our understanding of how the architectural environment can increase autonomy and independence, introduce therapy through design, mitigate scale, invite the family, support community services, fit into the surrounding context, and nurture social exchange and friendship formation.

Case studies that appear in Chapter Five are marked with an asterisk ✱

THE NETHERLANDS

1. Anton Pieckhöfje
Anijstraat 1
2034 ML Haarlem
Hans Houweling, M.D., Medical Director

Interesting two-story building with family housing on the upper floor and six small group homes for dementia victims (thirty-six residents) on the first floor. The group homes share a common walkway that links them around an open courtyard.

✱ **2. Bergzicht**
Alexberg 2
4800 DG Breda
Jacques Smit, Housing Director

A four-story enclosed street located adjacent to neighborhood stores. It provides limited services to residents and older people in the neighborhood. Bridges and unit entries are social meeting places.

3. de Boogerd
1-87 Prieelstraat
1628 Lt Hoorn n.h.
L. P. Kiunder, Resident

A larger eighty-seven unit communal experiment involving older people in voluntary efforts at self-help. Residents are members of ten task-oriented committees. They manage building maintenance and small groups prepare meals together in two communal kitchens and dining rooms.

4. de Drie Hoven
Louis Bouwmeestraat 377
1065 NS Amsterdam
D.H.R. Boyer, Director

A large geriatric hospital. This was architect Herman Hertzberger's first internationally recognized elder care facility. The two-story atrium acts as a "town center" for residents and staff, and is very effective. The "brown café" here is picturesque and popular.

5. de Kiekendief
Kolkplein 1
1315 GW Almere-Stad
Lammert Meyer, Head Nurse

A large, light-filled four-story central atrium with units on two sides is located adjacent to a shopping mall. A cafeteria and grocery store activate the atrium floor. The carpeted flat roof of these two spaces are lounges used as perches for residents that overlook the atrium below.

6. de Klinker
Borgenstraat 45
1053 PB Amsterdam
H. Moeskops, Administrator

An urban seventy-one-unit service house located adjacent to a busy market street in Amsterdam. One program is oriented toward taking care of forty older people in the surrounding neighborhood. Several innovative programs are targeted toward home repair and health and physical security.

7. de Kortenaer
Kortenaerstraat 7S
5703 En Helmond
E.V. Moorsel, Director

A steunpunt project with a long, straight, enclosed two-story street is a beautiful lightfilled atrium. A semicircular dome over the street gives it a tall, spacious scale and introduces plenty of light for plant growth.

8. de Opmaat
van Goghlan 2
2681 UA Monster
Ina Mouris, Director

This thirty-two-unit project adjacent to a health center provides outreach help to fifteen neighborhood units. Common spaces are popular with neighborhood residents and are managed by volunteers. The project also has four respite care short stay units.

9. de Overloop
Boogstraat 1
1353 BE Almere-Haven
Diana van Gaaven, Head Nurse

This project by architect Herman Hertzberger has received international attention. Note the clustered unit entries with glass windows to the corridor and the large light-filled central activity atrium. The controversial morgue/chapel on the top floor is an expressive curved mass visible from the street.

10. de Westerweeren
Burger Huybrechtstraat 66
DB Bergambacht
Caroline van Jaarsveld, Director

A thirty-unit experimental project in a rural area, located adjacent to thirty units of independent housing, the scale and commitment to caring for the frail in a residential environment is noteworthy. Canal and individual outdoor gardens are powerful site features.

11. Flesseman Center
Nieumarket 77
1011 MA Amsterdam
Anita Kinebanian, Manager

Located adjacent to historic Nieumarket Square in Amsterdam, this adaptive remodel of a historic "New Amsterdam"-style building creates a friendly courtyard. Erkers along the street edge give it a strong visual connection to street activity.

12. Höfje van Staats
Jansweg 39-1
Haarlem
S.C. Goetemelk, Contact Person

A beautifully restored höfje originally constructed in 1731, with a calm and peaceful courtyard, in the middle of the city. Remodeled units are two stories. Some are outfitted with chair lifts to accommodate stroke victims. Unit interiors are adaptively remodeled to fit modern life to the historic context.

13. Humanitas
Hennepstraat 4
7552 DN Hengelo
E. Koldewgn, Administrator

A forty-nine-unit atrium addition to an old building. The atrium is noisy and a bit stark. The nursing station with windows overlooking the atrium is in a dominant location.

*** 14. Jan van der Ploeg**
Hooglandstraat 67
3036 PD Rotterdam
Elly Ham, Project Leader

A four-story garden atrium on a tight urban renewal site. A bar and meal services program run by volunteers is also open to neighbors. A curved balcony decorated with hanging red geraniums gives residents a place to sit adjacent to their unit entry doors.

15. Kruistraat
14–128 Kruistraat St
3581 G.K. Utrecht
T.C. Kuiper, Architect

A modern project sensitively placed into an old area of Utrecht. A *Y*-shaped atrium and adjoining outside garden provide encouragement for walking. A glass serre (balcony sun room) is an attractive design feature for residents that functions either as an interior or exterior space.

16. Moerwijk
Twickelstraat 120
2539 RB Den Haag
G.M.M. Gronk, Director

A skillful remodeling of an older facility. A sloping greenhouse garden atrium at one end of a Y-shaped corridor is adjacent to the entry. Canals are intertwined with building forms.

17. Nieuw Doddendaal
Parkdwarsstraat 34
6511 DL Nijmegan
K. Schanzieh, Manager

This older home for the aged with small units was remodeled into a project with larger units that encourage independence. Located near the central part of the old city, the outside was left untouched and fits into the neighborhood well.

18. St. Bartolomeus Gasthuis
Lange Smeestraat 40
3511 PZ Utrecht
A. Schrijver, Administrator

Opened originally in 1367, this is one of the oldest continuously operated homes for the aged in the Netherlands. Located near the old canal in Utrecht, the scheme is organized around a landscaped courtyard. The original foundation director's room is beautifully preserved.

19. Stichtingete Bouwacher
Leidseweg 140
3533 HN Utrecht
Ron Van Aalst, Developer

Skillfully reuses an old 1870 country house as the community building. Semicircular entry drive establishes the unifying form for 111 new housing units. Units overlook an adjacent park and generous landscape areas.

20. Vereniging Anders Wonen
vor im Ouderen
Madame Curieplein 86
4834 XS Breda
Mrs. Pruyser, Resident

Interesting three-story, fourteen-unit communal living apartment. Contains a central glass skylight that daylights the stair and elevator. Windows to units introduce borrowed light from the atrium. Residents organized to build the project and moved in after construction.

21. Verzorgingshuis de Gooyer
von Zesenstraat + Dapperstraat
1093 BJ Amsterdam
Martin Spyker, Director

Service house on an urban site adjacent to a busy market street. Mixed retail use on the first floor completes the streetscape. A group home for dementia victims and voluntary technical assistance from a local nursing home allow management to keep most residents here until they die.

22. WZV Anholtskamp
A Ten Hovestraat 1
7475 CZ Markelo
A. Otten, Manager

Located in a small town, this building shares its common space with community organizations. Committed to keeping older local residents, people in the area, nursing patients, and home for the aged residents mixed in a successful hybrid scheme.

23. Zonnetrap
Molenvliet 572
Rotterdam
O. De Bot, Director

This large 180-unit, 9-story atrium building is constructed with units that step back on the south side to maximize sun exposure. Interesting dining room alcoves punch into the corridor balcony, giving residents a view of the activities below.

24. Dr. W. Drees
Morsestraat 19
2517 PM Den Haag
Wim van Schalk, Managing Director

A 119-unit, 5-story project serving a frail nursing home population in residential units. A large four-story atrium is openable during good weather. It daylights an activity center below which is open to the neighborhood.

DENMARK

1. Aktivitetscentret Baunbo
Skolegade 27 Lunde
6830 Nr. Nebel
Dagny Madsen, Director

A small rural twenty-six-unit housing project added to an old home for the aged that was in turn remodeled for community services. A one-story enclosed street links units together and is used for socializing. It has been loaned to high school students as a party venue for prom night.

2. Degneparken
Degneparken 15
4293 Dianalund
Kjeld Sørunsen, Technician

A small-town thirty-seven-unit service house that provides extensive health, therapy, and respite care to community members and older people living here. A two-story atrium provides an open plan community area with good natural light. It provides a range of recreational and therapeutic activities for residents and neighbors.

3. Dronning Anne-Marie Centret
Solbjerg Have 7
2000 Frederiksberg
Niels Gjarstrup, Managing Director

A nursing home (96 units) site-planned with a large social housing project (412 units). The nursing home has developed some intriguing policies for decentralized management. Older residents serve on the personnel committee and meet with staff on a biweekly basis to discuss ward policies and programs.

4. Frederiksbroën Aldrecenter
Georgsgade 61
5000 Odense C
Jhonny Krisdiansen, Director

This twenty-eight-unit project contains an interesting unit kitchen design. The kitchen preparation space juts into the corridor with windows that give each entry its own identity and a view of the corridor. Sidewalks that link the housing to local stores encourage residents to walk.

5. Fyensgadecentret
Fyensgade 25
9000 Ålborg
Soren Friis, Architect

A service house with a very expressive pyramid-shaped dining room ceiling is open to residents and neighbors. Interesting greenhouse alcoves face the street. Units are large with full kitchens.

6. Gulkrögcentret
Gulkrög 9
7100 Vieje
Anne Marie Henriksen, Physical Therapist

This project is master-planned to encourage pedestrian through traffic. Generous community facilities are shared with neighborhood residents. Social areas and lounges are designed to overlook sidewalks and plazas used by employees from a nearby hospital that cut through the site to catch a train on the other side of the site.

7. Lokalcenter Eskegården
Byagervej 115
8330 Beder
Karen Bulow, Occupational Therapist

A small one-story twenty-four-unit project with a large community services effort devoted to keeping older people independent in the surrounding neighborhood. Two hundred meals are delivered daily to nearby residents in an extensive meals-on-wheels project.

8. Lokalcenter Lille Gläsny
Kanslergade 12
5000 Odense C
Hamme Jorgensen, Architect

A new service house with a single-loaded U-shaped open balcony corridor. It is linked to a small service center. The forty-five units here follow the Danish standard for size (650–700 square feet) and are quite spacious.

9. Lokalcenter Skelager
Skelagervej 33
8200 Åarhus N
Jette Siggaåd, Director

A group living arrangement that challenges older residents to contribute as much as they can to the operation of the household. Facilitators organize tasks that are carried out by residents. Home help is provided for those who are not physically capable of doing things for themselves.

*** 10. Nybodergaården**
Kronprinsenssgade 61
1306 Kobenhavn K
Lisbeth Andersen, Manager

A forty-eight-unit, three-story atrium nursing home designed to conform to the shape of sixteenth-century housing in the Nyboder district of Copenhagen. Wide balconies allow residents to place furniture here that overlooks activities below. A neighborhood program cares for six off-site residents.

11. Omsorgscentret Egegården
Klausdalsbrovej 213
2860 Søborg
Lone Streibig, Administrator

A 112-bed nursing home with an open linear community building. Services and activities are arrayed along a main corridor. Clear glass transitions between the common building and residential wings provide places for residents to socialize, exercise, and read.

12. Pindstrup Centret
Perlevej 3
8550 Ryomgård
Olvf Ravn, Director

A summer camp for children (with a range of disabilities) that has pursued a number of interesting ideas regarding spaces for the blind, deaf, and developmentally disabled. Garden features and a handicapped-accessible fountain in the garden are particularly noteworthy.

13. Plejecentret Munke Mose
Allegade 94
5000 Odense C
Jane Hvas, Director

This fifty-unit project has used extensive single-loaded corridors to daylight circulation. It is adjacent to a park with views from the upper units. The courtyard in the center of the building has a sculpture and fountain from the older home it replaced.

14. Rosenborg Centret
Rosengade 1
1309 Kopenhavn K
Sif Galagain, Assistant Manager

A nursing home in Copenhagen with a courtyard and enclosed atriums (Altenwohnungen in Kopenhagen, 1979). The atrium is covered by a pyramid-shaped skylight and is used for morning and afternoon coffee. It fits well into the surrounding context.

15. Rygårdscentret
Niels Andersens vej 22
2900 Hellerup
Jannie Urban Hansen, Physical Therapist

An eighty-seven-unit service house with a substantial community services program. The outdoor courtyard is used for a range of physical therapy programs. Extensive use of turf block allows automobile access to most units.

16. Strandlund
Strandvejen 146
2920 Charlottenlund
Borge Koll, Director

An equity project in a park adjacent to the Baltic Sea that separates Denmark from Sweden. Housing is clustered in rows that overlook a beautiful park on one side and a combined pedestrian and vehicular turf block path on the other side.

SWEDEN

1. **Aspens Servicehus**
Bartotegatan 1
582 22 Linköping
Solveig Bödroth, Administrator

Older buildings located on the edge of a heavily trafficked pedestrian street were adaptively remodeled by preserving the scale and detail of these old structures. Children in a day-care program on the same site eat lunch with seniors in the service house restaurant. There are 109 units in several two- and three-story buildings.

2. **Atrium House for Pensioners**
Carlslund Park
194 22 Upplands Väsby
Helena Svensson, Information Officer

Six small attached dwellings cluster around a pyramid-shaped enclosed atrium that protects this space. The atrium area is carved into several different domains. Some spaces are communal and some are semiprivate extensions of resident dwelling units.

3. **Baltzargården**
Eenatriska Eliv Lasavettet
591 85 Motala
Barbro Sundelius, Director

A small group home for dementia residents that has been the basis for a book about the group home movement and the older people who live here. Motala Hospital has been actively involved in pursuing alternative programs for keeping older people in community settings and out of the hospital.

4. **BNF Kranen**
Oskarsrogatan 9
Solna
Rene Erland, HSB Information Chief

A 115-unit condominium project sponsored by one of Sweden's largest cooperative developers. The building is designed for the project population to age in place. Common rooms are set aside for home care workers who will be utilized as the need for home care service increases in the next decade.

5. **Fabriken**
Östra Storgatan 109
551 11 Jönköping
Anita Andersson, Administrator

A six-story project that mixes 179 family and elderly units. The building faces a busy commercial street. At one end of the project is a convenience store and at the other end is a shop that sells items created in the project's sheltered workshop.

6. **Fargärdet**
Breviksgatan 15
575 39 Eksjö
Bo Unmark, Housing Director

This small group home for eight residents with dementia is located next to a large hospital and adjacent to a grocery store in a small town. With attendant assistance, residents walk to the store and shop for food. Sloped roofs and a complex plan give the house a residential look from the street.

7. **Gruppboënde Orustgatan 16**
414 74 Göteborg
Elisabeth Andersson, Manager

This group home for six dementia residents is located on the top floor of a five-story building. This new building has been added to an older suburb that contains a number of multistory housing blocks. Many of the residents of this post-World War II suburb are older.

8. **Gruppboëndet Lönngården**
Granviksgatan 8
571 41 Nässjö
Marle Louise Jonsson, Director

This small group home for eight dementia residents is designed in conjunction with family housing in a small fishing village. An outdoor garden and patio deck provide places for residents to spend time outside. Some residents regularly visit the lake to fish with attendants.

9. **Gruppboëndet Strand**
Skeppargatan 15
671 30 Arvika
Barbro Wilson, Director

L-shaped group homes, located in a cluster plan, are two stories in height. Families live on the second floors and six physically or mentally frail older residents live below. Sloped ceilings and the extensive use of wood paneling give units a residential appearance.

10. Grupphemen Solbacken
Bergsmansvag 10–12
430 90 Öckerö
Rut Ludvigsson, Director

Two eight-unit group homes for dementia residents are located on a small fishing island. A spacious kitchen makes it an effective setting to involve older residents in meal preparation activities. Locating two homes together allows for economies at night because only one person is needed to monitor both settings.

*** 11. Hasselknuten**
Hasselknuten 5 A-F
444 44 Stenungsund
Inga Britt Johansson, Director

This dementia home for six residents is located on the first floor of a two-story L-shaped housing block, which also accommodates fifteen families. The project has a fenced outdoor garden and uses clear glass partitions between major common rooms to help orient confused older patients.

12. Kullëngens Sjukhem
Kullëngsvägan 1
694 00 Hallsberg
Julian Kimber, Head Nurse

A small courtyard and two small group eating arrangements for six persons each are the basis for this remodeling of a larger institution into a series of smaller wards. The new environment gives individuals a better chance to know one another. Creative murals decorate walls and columns in one courtyard cluster.

13. Kvarteret Karl XI
Källegatan 3
302 43 Halmstad
Stellen Eriksson, Architect

This service house has a glass corridor that runs through the center of the first floor linking a number of activity spaces. The U-shaped courtyard contains a portion of the medieval city wall that originally enveloped the city. In one housing block, a single-loaded enclosed glass corridor has encouraged residents to creatively "capture" this space as an extension of their dwelling unit.

14. Lundagården
Lundagårdsvägen
533 72 Lundsbrunn
Benit Bethlsson, Administrator

This local nursing home is designed around a glass-covered atrium that can be opened in good weather by mechanically sliding the roof to one side. Extensive use of wood materials in the floor, ceilings, and residential light fixtures give the setting a noninstitutional feel.

15. Mårtensund Servicehus
Brunnsgatan 11B
223 60 Lund
Sun Britt Lingren, Director

This large service house project utilizes separate floors for community and resident dining services. It uses extensive light-colored birchwood in units and along the corridors. The project also contains a swimming pool.

16. Rio Servicehus
Sandhamnsgatan 6 & 8
115 40 Stockholm
Sonny Melin, Assistant

This project, originally designed as a multi-tower communal housing experiment, was remodeled to allow residents more freedom of choice. The dining room is open to families and workers in the surrounding community, giving it the feeling of a neighborhood café rather than a dining room for the elderly.

17. Runby Servicehus
Runby Torg 9
194 40 Upplands Väsby
Birgitta Pettersson, Director

An interesting mixed-use building located over a neighborhood shopping center. The first-floor space contains doctors' offices and a library for the community. A large sloped atrium daylights and protects interior spaces.

18. Servicehuset Hornstull
Lignagatan 6
117 34 Stockholm
Agneta Birol, Administrator

A U-shaped courtyard building of six and seven stories opens onto an adjacent park (ServiceHuset Hornstull, Stockholm, 1985). This urban project combines a nursing home, community services, family housing, and housing for the frail. The use of a sloped roof and other residential details give this tall building the feeling that it fits into the housing sector of this part of Stockholm.

19. Skinnarvikens Servicehus
Heleneborgsgatan 2A-F
117 32 Stockholm
Margareth Lundgren, Administrator

A 165-unit project located on a steep Stockholm hillside. Six point towers are linked together at the base by a large continuous ground floor that contains a range of activities. Units have good views and elevator access to a range of stimulating activities and services.

20. Snöstorps Servicehus
Ljungbyvägen 28
302 Halmstad
Stina Åsen, Leader

This small eight-unit service house is sited to overlook the rolling countryside, with a beautiful garden defined by the U-shaped configuration of a one-story building. A south-facing service house porch invites sun in the morning.

21. Solgård Nursing Home
Storgattan 40
51 400 Tranemo
Anette Åkesson, Director

A good example of a local nursing home design that clusters four resident units around a sitting room and sixteen units around dining rooms. Courtyard designs are used to maximize views and access to outdoor spaces. An old Swedish ceramic fireplace is used to give the dining room character and warmth.

22. T-1
Överstegatan 22
581 03 Linköping
Inga Lena Perssen, Administrator

This new development is on the grounds of an old military base. The twenty-three-unit service house contains a gymnasium, meeting room, and crafts room designed to be shared by the older people, middle school children, and neighborhood residents. Young people also take meals in the service house restaurant.

23. Tornhuset
Åvägen 20
412 51 Göteborg
Barbro Thor, Manager

A six-story, 121-unit urban project located on an old 1923 site used originally as a soap factory. The administration building with the original clock tower was saved and redesignated for use as a community services building. The project surrounds and protects a garden on the inside of a courtyard.

24. Vastersol
Fabriksgatan 17
Jönköping
Anna Brita Nilsson, Director

This older hospital was recently remodeled for nursing purposes. Each floor is treated like a separate ward and residents eat with staff around small dining room tables. The wide double-loaded corridors have been decorated with antiques to make the environment more friendly.

25. Vickelbygården
Sågvagen 10 Box 112
617 00 Skärblacka
Lars Selevik, Administrator

This Swedish nursing home is modeled around Danish ideas. Small decentralized wards are wrapped around well-landscaped courtyards. Each residential unit has access to its own private patio space.

FINLAND

1. **Apian Palvelukeskus**
   Kangaskan 21
   376 00 Valkeakoski
   Sinikka Paty, Leader

This project is centered around a large U-shaped courtyard opening onto an urban plaza that links the project to the surrounding urban context. A glass observation space on top of the building is lighted at night, forming a highly visible beacon.

2. **Brahenpuiston Asuintalo**
   Porvoonkatulo
   005 10 Helsinki
   Kaarina Sainio, Director

This fifty-five-unit, four-story urban project is a mixed-use composition that combines first-floor retail with a health center, child-care center, and housing for the elderly. Approximately half the apartments overlook the internal courtyard, where a children's playground is located.

3. **Kotikallio Service Center**
   Kyläkirkontie 6-10E
   0037 Helsinki
   Paula Heikkinen, Administrator

This four-story service house combines extensive physical therapy facilities with a swimming pool and a thirty-five-unit inpatient rehabilitation hospital. One hundred and forty housing units clustered in the surrounding neighborhood use this building.

* 4. **Kuuselan Palvelukoti**
   Nuolialantie 46
   339 00 Tampere
   Anni Lvokkala, Director

This garden atrium project has a swimming pool on the lower level that overlooks the central atrium. Two dementia group homes for four residents each are located on the fourth floor. Community services like meals-on-wheels and emergency response are coordinated out of this building.

5. **Old Peoples Home and Health Center**
   Kuusitie 10–18
   121 00 Oitti
   Päivi Terävä, Social Services Director

This dramatic forty-two-unit project resulted from an architectural competition. The program combines housing and an outpatient health center. Housing is centered around four village clusters grouped around a two-story pyramidal skylight. A continuous corridor broken in several places links the separate unit cluster "villages."

6. **Palvelutalo Esikko**
   Uitta Montie 7
   208 10 Turku
   Markku Jyväs, Director

This equity project has a cost-sharing agreement with the local municipality that allows a swimming pool and restaurant to be shared with neighborhood residents. The central four-story atrium has a wall of viney plant materials adjacent to balcony corridors.

7. **Riihimäen Vanhainkoti**
   Kontiontie 73
   111 00 Riihmäki
   Päivi Pelkonen, Director

This fifty-five-unit U-shaped courtyard scheme is an addition to an older building (Riihimäen Vanhainkoti, 1985). On the top floor is an unusual clerestory roof element that brings light into double-loaded corridors and allows borrowed light to reach the dining room, which gets light from two sides.

8. **Vanhainkoti-Parvakeskus Himmeli**
   Palokunnantie 39
   281 30 Pori
   Irma Roininen, Director

This home for the aged contains an interesting corridor configuration that terminates in a two-story fireplace. A child-care center attached to the home has introduced joint programs for children and older people. The large well-landscaped site provides opportunities for exercise.

9. **Viherkoti**
   Kuusiniemi 13–15
   027 10 Espoo
   Maire Koski, Head Nurse

This fifty-unit nursing home contains a three-story atrium. Half of the atrium space is devoted to physical therapy and exercise. A garden, adjacent to the atrium, allows large areas to be combined and devoted to events that involve residents with families.

NORWAY

1. Aldersbustadane Service Sentret
Sund Bø Haugen
3841 Flatdal
June Olsen, Leader

This small service house accommodates four units in a plan centering around a communal room where meals are taken. Small sitting areas created by the unit configuration and a covered walkway that connects them make the outdoor common areas comfortable and usable.

2. Boliger Pa Banken
Nedre Bankegatan
1750 Halden
Thinh Huu Nguyen, Resident

This is a beautifully restored eighteenth-century courtyard housing arrangement of twenty-three units located in the middle of a small town. The central courtyard contains a mixture of private and communal spaces and is a successful social space. It is an age-integrated setting with a majority of older residents.

3. Forsmannsenteret
Dronningsgatan 24
3200 Sandefjord
Solveig Walloe, Resident

This active service center contains a number of special-use recreation spaces for older people who live in the surrounding neighborhood. The building facade creates a strong urban edge, with retail uses on the first floor. A courtyard greenhouse and restaurant overlooking the south edge of the site provide attractive places for residents to sit.

4. Furubakken Service Boliger
Ovre Torggatan 2
2800 Gjøvik
Liv Botten Bjørklund, Director

Located on the edge of a hill in a medium-sized town, this forty-five-unit service house has a day-care program for community members, which is also popular with project residents (Furubakken Service Boliger, 1990).

5. Lårdal Alderspensjonat
3860 Høydalsmo
Güdrün Neverdalen, Leader

A small eight-unit service house located in a rural village blends remodeled housing with new construction. A library, small dining room, kitchen, and recreation space are shared with residents.

*** 6. Lesjatun**
2665 Lesja
Hans Hesthagen, Social Services Director

A small service house of fifteen units on the edge of a dramatic valley represents the best in the Norwegian tradition of building in wood. Units have rustic wood walls and ceilings. The south-facing balcony provides views of the surrounding landscape.

7. Midtløkken Boog Servicesenter
Kong Oscarga-tan 15
3100 Tønsberg
Ellen Otterstad, Director

This fifty-seven-unit service house contains a pool, extensive community space, and a restaurant below an air-rights equity condominium. The U-shaped courtyard opens onto a park, which encourages residents to walk to the city center a few blocks away.

8. Raufosstun Eldresenter
Severin Olsens Vei 15
2830 Raufoss
Line Kjosbakken, Director

This thirty-two-unit home for the aged is planned around small courtyard tuns that connect rear patio areas. It is painted a traditional bright red, which contrasts with the green landscape of the summer and the white snow covering of the winter.

9. Søreidtunet Eldresentret
Søreidtunet 2
5060 Søreidgrend
Aslang Hesjedal, Leader

A thirty-nine-unit service house, which contains doctors' offices and a post office. Housing located above is splayed against a sloped hillside. Units share a narrow forecourt in a space reminiscent of the long thin Hanseatic buildings on the old Bergen wharf.

10. Sportsvenien Borettslag
Lyseagan 50
0383 Oslo 3
Reidar Jacobsen, Resident Manager

This fifty-four-unit semicircular-shaped cooperative is for younger elderly. A single-loaded balcony corridor with storage spaces and sitting alcoves has created places for residents to interact with one another. Private balconies are carved out of the curved form on one side.

11. Ulvøya Eldreboliger
Pans Vei 1–3
0139 Ulvøya Oslo 1
Wilhelimae Warness, Resident

This small sixteen-unit village is located on a small island near Oslo. A main arcade links porches and entry spaces together. A common room has been loaned to the district doctor who sees patients here every week, including children and families from throughout the island.

UNITED STATES OF AMERICA

1. Annie Maxim House
700 North Avenue
Rochester, MA 02770
Karen Greene, Director

This horseshoe-shaped congregate house is linked by an enclosed porch that allows residents to observe activities on the site. A central commons area has extensive opportunities for reading, socializing, and playing games. The scale of the twelve-unit setting and its rural location add to its charm.

2. The Argyle
4115 W. 38th Ave
Denver, CO 80212
Ann Brown, Administrator

This remodeled facility has restored exterior facade elements and interior casework treatments from a historic building constructed in 1899. The overall look is residential and historic in character. A patio has been created that is picturesque.

3. Arizona Senior Homes
1041 S. LeBaron
Mesa, AR 85017
Stan Hosac, Developer

In this project, six separate houses accommodate ten persons each, and are located on two sides of a dead-end street. The project fits perfectly into the surrounding neighborhood using the conventional house as the camouflage. It appears to be a continuation of existing housing along the street.

4. John Bertram House
29 Washington Square
Salem, MA 01970
William Carney, Administrator

This project in Salem is a beautifully remodeled nineteenth-century home across from a large park. The dining rooms and common spaces on the first floor have been carefully restored. The country kitchen is designed around an alcove with windows that let residents preview the space.

5. Brighton Gardens
5620 Wesleyan Drive
Virginia Beach, VA 23455
John Prose, Manager

This prototype developed by the Marriott Corporation is a courtyard housing arrangement that combines assisted living and nursing care. Residential interiors and informal spaces for residents are popular. The units are small but have a number of thoughtful features.

✱ 6. Captain Clarence Eldridge House
30 Pine Street
Hyannis, MA 02601
Bonnie Goodwin, Administrator

This congregate house designed to accommodate twenty people is an expansion of a nineteenth-century sea captain's dwelling on Cape Cod. A central atrium and dwelling units with double-hung windows and Dutch doors open the unit interiors to natural light. Residents here know one another and have created important instrumental helping exchanges.

7. Eaton Terrace II
323 South Eaton
Lakewood, CO 80226
Dean Painter, Administrator

This mid-rise six-story building is attached to a nonprofit subsidized independent housing project. Residents of the subsidized housing take personal care service here rather than move to a nursing home. Units have accessible kitchens and bathrooms that are handicapped-adaptable. A tiled alcove on the outside edge of the living room provides a place for plants.

8. Elder Homestead
11400 4th St. North
Minnetonka, MN 55343
Jan Stenzel, Manager

This small-scale twenty-nine-unit project was designed to resemble a rambling rural Minnesota farmhouse. Four individual units share a cluster parlor. Common spaces are in scale with the residential nature of the building. A three-season enclosed porch allows residents to view the surrounding street.

9. Motion Picture and Television Fund Country Home and Hospital
23388 Mulholland Drive
Woodland Hills, CA 91364
William Haug, Executive Director

The Alzheimer's unit here has a beautiful garden which is linked to the facility and allows residents living here to experience several different ecologies. A civic park, running brook, and several different sunny and shady places to sit adjacent to a looped pathway make it interesting for the residents.

10. Mount San Antonio Gardens Lodge
900 E. Harrison
Pomona, CA 91767
Ted Radamaker, Administrator

The assisted living unit of this large CCRC has remodeled its multipurpose room into a gym, where residents can use physical therapy equipment. A rigorous program of therapy keeps as many residents as possible from moving to the nursing unit. The U-shaped building is situated around a large oak tree, which provides shade and focus to the courtyard.

11. Rackleff House
655 SW 13th Ave
Canby, OR 97013
Keren Wilson, Director

This twenty-five-unit single-story courtyard arrangement in a small rural Oregon town is on a site with large mature trees. The corridor that links units is double-loaded on two sides and single-loaded on two sides, opening the project to views of the courtyard.

*** 12. Rosewood Estate**
2750 N. Victoria St.
Roseville, MN 55113
Alan Black, Administrator

This sixty-eight-unit alternative nursing home is designed around a home care model that maximizes independence. A physical massing configuration segments the building into three pieces giving it the look of a mansion house from the street.

13. Sunrise Retirement Home of Bluemont Park
5900 Wilson Boulevard
Arlington, VA 22205
Mary Brickel, Administrator

This project of approximately 150 units is broken into three relatively self-contained buildings, which appear separate from the entrance courtyard. They are linked below grade by a service corridor that facilitates delivery of food and laundry services.

*** 14. Sunrise Retirement Home of Fairfax**
9207 Arlington Blvd
Fairfax, VA 22031
David Peete, Administrator

This three-story Victorian house accommodates forty-seven units in an L-shaped building. The treatment of the exterior and interior design is consistent with ideas about residential imagery. Spaces are small in scale and add to the feeling that it is a home and not an institution.

*** 15. Woodside Place of Oakmont**
1215 Hulton Rd
Oakmont, PA 15139
Arlene McGannon, Administrator

This one-story, thirty-unit residential facility for Alzheimer's residents allows patients free reign to wander. The massing involves three small cottages linked through a wandering path to a large multipurpose room which resembles a barn. The Shaker details used for interior treatments give it a homey feeling.

APPENDIX B

Glossary of Housing Concepts

INTRODUCTION

Each culture has invented descriptions for the housing types that constitute its contribution to the choices available to older frail people. Although not exhaustive, the following thirteen building concepts are represented in the examples chosen for this book. Within these definitions are examples of how the physical and service environment can be integrated. They represent ideas about urbanism, care philosophies, community building, and the provision of privacy.

1. **Point of Support Housing** Dutch housing model designed to provide limited services to residents and older people living in the surrounding community. Dwelling units are connected to common spaces by a protected and enclosed pathway.

2. **Home for the Aged** Designed for older frail with personal care needs. Many new European models are designed to allow residents to age in place instead of being moved to a nursing home.

3. **Service House** Common to all five European cultures that were the subject of this study. The service house combines housing for the frail with a range of other community services, including a restaurant, multipurpose senior center, adult day care, home care services, occupational therapy, and physical therapy. Services are available to residents living in the neighborhood as well as to project residents.

4. **Local Nursing Home** Swedish prototype developed in the early 1980s with the intention of decentralizing large wards, creating better relationships between indoor and outdoor spaces, and reducing the scale of the building. Many are designed around courtyards with twelve to fifteen rooms in each cluster.

5. **Plegehem** Danish nursing home model focused around therapy and autonomy. Often includes a bar on the first floor or in the main lobby as a socializing catalyst. Decentralized and organized around small wards, residents eat in small groups around tables, family-style. Ideas about participatory governance are also well developed in many projects.

6. **Life-Span Housing** Norway's adaptable housing standards developed to allow elderly housing to serve the needs of frail persons as they age in place. Standards are enforced through Husbanken, the national housing bank.

7. **Communal Housing** These projects employ a range of different strategies that allow residents to collectively manage housing and services. Projects range from those in which communal participation in cleaning, cooking, and maintenance are optional to those organized around facilitators who assign tasks and monitor participation. In general, communal arrangements have been less effective dealing with vulnerable and frail populations.

8. **Group Home** Small group living arrangements for six to eight older dementia victims, located on floors of high-rise buildings, in purpose-built cottage housing or in low-rise age-mixed social housing schemes. Sometimes several projects are clustered together for economy of scale purposes. They employ a normal daily activity philosophy, using attendants to motivate residents to perform everyday activities in a small-scale noninstitutional environment.

9. **Lean-to Housing** A Dutch housing type often constructed near service houses, point of support housing, or homes for the aged. Residents living here lead independent lives but can use adjacent meal and personal care services. They are often electronically linked to service centers through an emergency call system. The housing combines independence with immediate access to services and emergency response.

10. **European Nursing Homes** These health care settings are for the most impaired older people and are similar to subacute hospitals in the United States. Residents in modern facilities have private single-bed accommodations. Community service models with outreach to the surrounding neighborhoods are common. Physical and occupational therapy is an important dimension of the lifestyle in these environments.

11. **Congregate House** A housing model that often involves a form of volunteer collective living. Residents in small homelike dwellings live independently but can agree to take meals together. The services provided include housekeeping assistance, meal preparation, and personal assistance with bathing. Projects stop short of providing twenty-four-hour personal care, which means that early dementia victims and those with night care needs are difficult to support.

12. **Assisted Living Housing** United States housing type often licensed under board and care or personal care regulations by state. The perceived and real limitations implied by licensure vary from state to state. Generally, health care services are limited and placement in a nursing home may be required if 24 hour health care supervision is needed. Residents typically receive three meals a day and are provided a range of personal care services including but not limited to bathing and toileting assistance, medication supervision, and help with dressing and grooming. Some progressive states are creating new regulations for assisted living that legalize a broader range of health care services placing it closer to skilled nursing care than board and care.

13. **Home-Care-Based Nursing** An unlicensed United States housing alternative to the conventional nursing home. Meals and housing are provided on a month-to-month lease similar to congregate housing. Medical and personal care needs are assessed by a home care agency and a plan devised that involves a combination of health and personal care support from family members and paid home care professionals.

BIBLIOGRAPHY

"Altenheim in Amsterdam." (1976). *Baumeister*, (2), pp. 132–36.

"Altenheim Aspen in Linköping." (1980). *Baumeister*, (11), pp. 1132–33.

"Altenwohnungen in Kopenhagen." (1979). *Baumeister*, (6), pp. 593–95.

"Apian Palvelutalo." (1988). *Arkkitehti*, 85, (3), pp. 67–9.

"Auf dem Polder." (1985). *Deutsche Bauzeitung, 119*, (11), pp. 17–20.

"De Drie Hoven Old People's Center." (1976). *The Architectural Review*, (2), pp. 74–7.

"Formannssenteret." (1989). *Byggekunst*, (6), pp. 422–25.

"Furubakken Serviceboliger." (1990). Oslo: Norges Byggforskningsinstitutt.

"Lesjatun Serviceboliger." (1990). Oslo: Norges Byggforskningsinstitutt.

"Livslopsboliger." (1985). Oslo: Norges Byggforskningsinstitutt.

"Lundagårdens Sjukhem, Lundsbrunn." (1984), *Arkitektur*, (1), pp. 12–15.

"Old People in Amsterdam." (1977). *Domus*, (569), April, pp. 11–13.

"Raufosstun." (1979). *Byggekunst*, (3), pp. 175–77.

"Residenza per anziani a Almere-Haven." (1984). *Casabella*, pp. 504–8.

"Riihimaen Vanhainkoti." (1985). *Arkkitehti*, 82, (5), pp. 62–7.

"Rygardcentret i Gentofte." (1981). *Arkitektur DK*, pp. 294–301.

"Servicehus Halmstad." (1986). *Arkitektur*, (3), 1986. pp. 38–41.

"ServiceHuset Hornstull, Stockholm." (1985). *Arkitektur*, (3), pp. 30–5.

"Strandlund, en boligbebyggelse for aeldre i Charlottenlund." (1983). *Landskab*, (5), pp. 97–9.

"Vanhakoiviston Vanhainkoti." (1989). *Arkkitehti*, 86, (8), pp. 35–40.

"Wooncentrum Nieuwmarkt Te Amsterdam." (1990). *Bouw*, (6), March 23, pp. 33–40.

AIA Foundation. (1985). *Design for Aging: An Architect's Guide.* Washington, D.C.: AIA Press.

Alexander, C., S. Ishikawa, and M. Silverstein. (1977). *A Pattern Language.* New York: Oxford University Press.

Almberg, C., and J. Paulsson. (1988). *Gruppboënde for aldersdementa.* Goteborg: Chalmers Division for Housing Design, R7.

_____. (1991). "Group Homes and Groups of Homes: Alternative Housing Concepts and their Application to Elderly People with Dementia in Sweden," in W. Preiser, J. Vischer, and E. White (eds.), *Design Intervention: Toward a More Humane Architecture.* New York: Van Nostrand Reinhold.

ASVVO. (1991). *Integratie Van Verzorging en Verpleging.* Amsterdam: ASVVO.

Beck-Friis, B. (1988). *At Home at Baltzargården.* Orebro: Bokforlaget Libris.

Beckman, M. (1976). *Building for Everyone.* Stockholm: Ministry of Housing and Physical Planning.

Benjamin, A., and R. Newcomer. (1986). "Board and Care Housing: An Analysis of State Differences." *Research on Aging, 8,* (3), pp. 388–406.

Beyer, H., and F. Nierstrasz. (1967). *Housing the Aged in Western Countries: Programs, Dwellings, Homes and Geriatric Facilities.* New York: Elsevier.

Boles, D. (1985). "Congregate Manor." *Progressive Architecture, 66,* (8), pp. 99–103.

Bull, G. (1987). *Boliger for Eldre.* Oslo: Bolig for Livet.

Bull, G., and I. Lise-Saglie. (1991). "Service Flats: An Alternative to Institutions in the Care of the Elderly." Oslo: Norwegian Building Research Institute.

Calkins, M. (1988). *Designing for Dementia: Planning Environments for the Elderly and Confused.* Owings Mills: National Health Publishing.

Carstens, D. (1985). *Site Planning and Design for the Elderly: Issues, Guidelines and Alternatives.* New York: Van Nostrand Reinhold.

Chappell, N. (1990). "Aging and Social Care," in R. Binstock, and L. George (eds.), *The Handbook of Aging and the Social Sciences.* San Diego: Academic Press.

Christoffersson, A., and R. Runbro. (1987). *Det Osynliga Folket.* Helsingborg: Carlssons Bokförlag.

Cohen, M., E. Tell, and S. Wallack. (1986). "The Lifetime Risks and Costs of Nursing Home Use Among the Elderly." *Medical Care, 24,* (12), pp. 1161–72.

Cohen, U., and K. Day. (1991). *Contemporary Environments for People with Dementia.* Milwaukee: Center for Architecture and Urban Planning Research.

Cohen, U., and J. Weisman. (1991). *Holding on to Home: Designing Environments for People with Dementia.* Baltimore: Johns Hopkins University Press.

Contemporary Long Term Care. (1992a). "1992 Interior Design Awards." *Contemporary Long Term Care, 15,* (5), pp. 33–54.

_____. (1992b). "1992 Architectural Design Awards." *Contemporary Long Term Care, 15,* (6), pp. 43–70.

Cooper, C. (1976). "Swimming for Senior Citizens." *Therapeutic Recreation Journal, 2,* pp. 50–54.

Crimmins, E., and D. Ingegneri. (1990). "Interaction and Living Arrangements of Older Parents and their Children: Past Trends, Present Determinants, Future Implications." *Research on Aging, 12,* (1), pp. 3–33.

Daatland, S. (1991). "Nordic Policies for Service and Care for the Elderly." Oslo: Norwegian Institute of Gerontology.

Det Hanseatiske Museum. (1982). *Bryggen: The Hanseatic Settlement in Bergen.* Bergen: Det Hanseatiske Museum.

Dobkin, L. (1989). *The Board and Care System: A Regulatory Jungle.* Consumer Affairs Program, Washington, D.C.: AARP.

Finnish Architectural Review. Hausjarvi Health Center and Old People's Home Design Competition, 2/87.

Franck, K., and S. Ahrentzen. (1990) *New Households, New Housing.* New York: Van Nostrand Reinhold.

Gaskie, M. (1988). "A Little Help: Housing for the Aging." *Architectural Record, 176,* (4), pp. 98–107.

Gehl, J. (1987). *Life Between Buildings: Using Public Space.* New York: Van Nostrand Reinhold.

Goldenberg, L. (1982). *Housing for the Elderly: New Trends in Europe.* New York: Garland Press.

Green, I., B. Fedewa, C. Johnston, W. Jackson, and H. Deardorff. (1975) *Housing for the Elderly: The Development and Design Process.* New York: Van Nostrand Reinhold.

Guralnik, J., M. Yanagishta, and E. Schneider. (1988). "Projecting the Older Population of the United States: Lessons From the Past and Prospects for the Future." *The Milbank Quarterly 66,* (2). pp. 283–308.

Hauglid, A. (1989). *Maihaugen: The Sandvig Collections.* Lillehamer: Mesna Trykk.

Heumann, L. F., and D. Boldy. (1982). *Housing for the Elderly: Policy Formulation in Europe and North America.* London: St. Martin's Press.

Hiatt, L. (1991). *Nursing Home Renovation Designed for Reform.* Boston: Butterworth Architecture.

Hoglund, D. (1985). *Housing for the Elderly: Privacy and Independence in Environments for the Aging.* New York: Van Nostrand Reinhold.

Howell, S. (1980). *Designing for Aging: Patterns for Use.* Cambridge: MIT Press.

Husbanken. (1987). *La Oss bo Midt i Livet!* Oslo: Husbanken.

Kacavas, J., D. Morrison, and M. Hurley. (1977). "The Use of Aqua-Therapy with Geriatric Patients." *American Corrective Therapy Journal,* March–April, pp. 52–9.

Kalymun, M. (1990). "Toward a Definition of Assisted-Living," in L. Pastalan, (ed.), *Optimizing Housing for the Elderly: Homes Not Houses*. New York: The Hayworth Press, Inc., pp. 97-132.

Kane, R., L. Illuston, R. Kane, and J. Nyman. (1990). *Meshing Services with Housing: Lessons from Adult Foster Care and Assisted Living in Oregon*. Minneapolis: University of Minnesota, Long-Term Care DECISIONS Resource Center.

Koncelik, J. (1976). *Designing the Open Nursing Home.* Stroudsburg: Dowden, Hutchinson, and Ross.

———. (1982). *Aging and the Product Environment.* Stroudsburg: Hutchinson and Ross.

Kuller, R. (1991). "Familiar Design Helps Dementia Patients Cope," in W. Preiser, J. Vischer, and E. White (eds.), *Design Intervention: Toward a More Humane Architecture.* New York: Van Nostrand Reinhold.

Lauvli, M. (1991). *Utbygging Av ServiceBoliger I Norge, Rapport 2.* Oslo: Norsk Gerontologisk Institutt.

Lauvli, M., and O. Guntvedt. (1988). *Service Boliger for Eldre, Rapport 3.* Oslo: Norsk Gerontologisk Institutt.

Lawton, M. P. (1975). *Planning and Managing Housing for the Elderly.* New York: John Wiley & Sons.

———. (1980). *Environment and Aging.* Monterey: Brooks/Cole Publishing.

Lawton, M. P., and L. Nahemow. (1973). "Ecology and the Aging Process," in C. Eisdorfer and M.P. Lawton (eds.), *Psychology of Adult Development and Aging.* Washington, D.C.: American Psychological Association.

Liebowitz, B., M. P. Lawton, and A. Waldman. (1979). "A prosthetically designed nursing home." *American Institute of Architects Journal, 68*, pp. 59-61.

Lindstrom, B. (1989). *Gode Boliger til Aeldre.* Copenhagen: Bybberiets Udviklingsrad.

Long Term Care National Resource Center at UCLA/USC. (1989). *Assisted Living Resource Guide.* Los Angeles: The Center.

Lundin, L., and B. Turner. (forthcoming). "Housing Policy for Frail Older Persons in Sweden," in J. Pynoos and P. Liebig (eds.), *Housing Policy for Frail Older Persons: International Perspectives and Prospects.* Philadelphia: Temple University Press.

McRae, J. (1975). *Elderly in the Environment: Northern Europe.* Gainesville: Center for Gerontological Studies, University of Florida.

Malmberg, B., and I. Oremark. (1991). *Enutvardering Av Fyra Gruppboënden For Alders-Dementa I Jönköping, Rapport #72.* Jönköping Institute for Gerontology.

MOCA. (1989). *Blueprints for Modern Living.* Cambridge: MIT Press.

Moore, C., G. Allen, and D. Lyndon. (1974). *The Place of Houses.* New York: Holt, Rinehart and Winston.

Morton, D. (1981). "Congregate Living." *Progressive Architecture, 62*, (8), pp. 64-8.

Myers, G. (1990). "Demography of Aging," in R. Binstock and L. George (eds.), *The Handbook of Aging and the Social Sciences.* San Diego: Academic Press.

Pensioner in Denmark. (1990). Copenhagen: Ministry of Social Affairs.

Preiser, F. (1978). *Facility Programming.* New York: McGraw-Hill.

Pynoos, J., and P. Liebig (eds.). (forthcoming). *Housing Policy for Frail Older Persons: International Perspectives and Prospects.* Philadelphia: Temple University Press.

Pynoos, J., and V. Regnier. (1991). "Improving Residential Environments for the Frail Elderly: Bridging the Gap Between Theory and Application," in J. Birren, J. Lubben, J. Rowe, and D. Deutchman (eds.), *The Concept and Measurement of Quality of Life in the Frail Elderly.* New York: Academic Press.

Raper, A., and A. Kalicki. (1988). *National Continuing Care Directory.* Glenview: Scott, Foresman and Co.

Rapoport, A. (1969). *House Form and Culture.* Englewood Cliffs: Prentice-Hall.

Raschko, B. (1982). *Housing Interiors for the Disabled and Elderly.* New York: Van Nostrand Reinhold.

Regnier, V. (1985). *Behavioral and Environmental Aspects of Outdoor Space Use in Housing for the Elderly.* Los Angeles: School of Architecture, Andrus Gerontology Center, University of Southern California.

———. (1992). *European Models of Assisted Living Housing for Mentally and Physically Frail Older People,* (Videotape). Los Angeles: National Eldercare Institute on Housing and Supportive Services, University of Southern California, Andrus Gerontology Center.

Regnier, V., J. Hamilton, and S. Yatabe. (1991). *Best Practices in Assisted Living: Innovations in Design, Management and Financing.* Los Angeles: The National Eldercare Institute on Housing and Supportive Services, University of Southern California.

Regnier, V., J. Hamilton, and S. Yatabe. (forthcoming). *Assisted Living for the Frail and Aged: Innovations in Design, Management and Financing.* New York: Columbia University Press.

Regnier, V., and J. Pynoos. (1987). *Housing the Aged: Design Directives and Policy Considerations.* New York: Elsevier.

Regnier, V., and J. Pynoos. (1992). "Environmental Interventions for Cognitively Impaired Older Persons," in J. Birren, B. Sloane, and G. Cohen (eds.), *Handbook of Mental Health and Aging*, Second Edition. New York: Academic Press.

Reinius, K. (1984). *The Elderly and their Environment: Research in Sweden.* Stockholm: Swedish Council for Building Research.

Rivlin, A., and J. Wiener. (1988). *Caring for the Disabled Elderly.* Washington, D.C.: The Brookings Institute.

Robinette, G. (1985). *Barrier-Free Exterior Design.* New York: Van Nostrand Reinhold.

Rose, E., and N. Bozeat. (1980). *Communal Facilities in Sheltered Housing.* Farnsborough: Saxon House.

Rosow, I. (1967). *Social Integration of the Aged.* New York: Free Press.

Rudovsky, B. (1969). *Streets for People: A Primer for Americans.* Garden City,: Doubleday and Co.

Rutherford, R., and A. Holst. (1963). *Architectural Designs: Homes for the Aged. The European Approach.* Peoria: Howard Company.

Rybczynski, W. (1989). *The Most Beautiful House in the World.* New York: Penguin Books.

Seip, D. (1989a). "First National Assisted Living Industry Survey." *Contemporary Long Term Care, 12*, (7), pp. 69–70.

_____. (1989b). "Free-Standing Assisted Living Trends." *Contemporary Long Term Care, 12*, (12), pp. 20, 22–3.

_____.(1989c). "Tallying the First National Assisted Living Survey." *Contemporary Long Term Care, 12*, (10), pp. 28, 30, 32–3.

_____. (1990). *The Survival Handbook for Developers of Assisted Living.* Boca Raton: The Seip Group.

SEV. (1991a). *Ouderen Wonen Anders in Breda.* Rotterdam: SEV.

_____. (1991b). *Wijksteunpunten voor Ouderen.* Rotterdam: Stuurgroep Experimenten Volkshuisvesting (SEV).

Soderstrom, B., and E. Viklund. (1986). *Housing Care and Service for Elderly and Old People: The Situation in Sweden.* Stockholm: The Swedish Ministry of Housing and Physical Planning.

Spirduso, W., and P. Gilliam-McRae. (1991). "Physical Activity and Quality of Life in the Frail Elderly," in J. Birren, J. Lubben, J. Rowe, and D. Deutchman (eds.), *The Concept and Measurement of Quality of Life in the Frail Elderly.* San Diego: Academic Press.

SPRI. (1979). *Lokala Sjukhem: Underlag for planering och projektering.* SPRI Rapport 12, Stockholm: Sjukvardens och socialvardens planerings-och rationaliseringsinstit.

____. (1980). *Lokala Sjukhem.* SPRI Rapport 35, Stockholm: Sjukvardens och socialvardens planerings-och rationaliseringsinstit.

____. (1989). *Vad kostar det att bygga om ett sjukhem?* SPRI Rapport 267, Stockholm: Sjukvardens och socialvardens planerings-och rationaliseringsinstit.

____. (1990). *Gruppboënde pa sjukhem.* SPRI Rapport 280, Stockholm: Sjukvardens och socialvardens planerings-och rationaliseringsinstit.

Steinfeld, E. (1987). "Adapting Housing for Older Disabled People," in V. Regnier and J. Pynoos (eds.), *Housing The Aged: Design Directives and Policy Considerations.* New York: Elsevier.

Stiching Wonen Amsterdam. (1987). *Young Dutch Architects 1987 Biennale: Housing for the Elderly in the City.* Amsterdam: Stiching Wonen.

Sundstrom, G. (1987). *Old Age Care in Sweden: Yesterday, Today. . . and Tomorrow?* Stockholm: The Swedish Institute.

Tilson, D. (1990). *Aging in Place: Supporting the Frail Elderly in Residential Environments.* Glenview: Scott, Foresman and Co.

Toneman, H. (1990). *Woongemeenschappen van Ouderen.* The Hague: Del Wel.

U.S. Bureau of the Census. (1983). "America in Transition: An Aging Society," by Cynthia M. Taeuber. *Current Population Reports* Series P-23, No. 128.

U.S. Bureau of the Census. (1984). "Projections of the Population of the United States, by Age, Sex and Race: 1983 to 2080," by Gregory Spencer. *Current Population Reports* Series P-25, No. 952.

U.S. Bureau of the Census. (1989). "Projections of the Population of the United States, by Age, Sex and Race: 1988 to 2080," by Gregory Spencer. *Current Population Reports* Series P-25, No. 1018.

U.S. Senate Special Committee on Aging. (1988). *Aging America: Trends and Projections.* Washington, D.C.: USDHHS.

U.S. Senate Select Committee on Aging. (1989). "Board and Care Homes in America: A National Tragedy." *Subcommittee on Health and Long Term Care.* Washington, D.C.: USGPO.

____. (1991). *Aging America: Trends and Projections.* Washington, D.C.: USDHHS.

Valins, M. (1988). *Housing for Elderly People: A Guide for Architects and Clients.* New York: Van Nostrand Reinhold.

Weal, F., and F. Weal. (1988). *Housing for the Elderly: Options and Design*. New York: Nichols Publishing.

Welch, P., V. Parker, and J. Zeisel. (1984). *Independence Through Interdependence*. Department of Elder Affairs, Commonwealth of Massachusetts.

Whyte, W. (1980). *The Social Life of Small Urban Spaces.* Washington, D.C.: Conservation Foundation.

Willcocks, I., S. Peace, and L. Kellaher. (1986). *Private Lives in Public Places.* London: Tavistock Press.

Wilson, K. B. (1990). "Assisted Living: The Merger of Housing and Long Term Care Services." *Long Term Care Advances*, 1, p. 208.

WVC. (1989). *Matching the Care with the Needs of the Elderly*. Ministry of Welfare, Health, and Culture, Rijswijk.

Zeisel, J. (1981). *Inquiry by Design.* Monterey: Brooks/Cole Publishing.

Zeisel, J., G. Epp, and S. Demos. (1977). *Low-Rise Housing for Elderly People: Behavioral Criteria for Design.* Washington, D.C.: U.S. Government Printing Office, #HUD-483, September.

Zeisel J., P. Welch, G. Epp, and S. Demos. (1983). *Mid-Rise Elevator Housing for Older People.* Boston: Building Diagnostics.

Zoet, M., and H. Klieverik. (1990). *Woningraad*, (4), pp. 13–5.

INDEX